Fieldwork for Healthcare:

Guidance for Investigating Human Factors in
Computing Systems

Synthesis Lectures on Assistive, Rehabilitative, and Health-Preserving Technologies

Editor
Ron Baecker, *University of Toronto*

Advances in medicine allow us to live longer, despite the assaults on our bodies from war, environmental damage, and natural disasters. The result is that many of us survive for years or decades with increasing difficulties in tasks such as seeing, hearing, moving, planning, remembering, and communicating.

This series provides current state-of-the-art overviews of key topics in the burgeoning field of assistive technologies. We take a broad view of this field, giving attention not only to prosthetics that compensate for impaired capabilities, but to methods for rehabilitating or restoring function, as well as protective interventions that enable individuals to be healthy for longer periods of time throughout the lifespan. Our emphasis is in the role of information and communications technologies in prosthetics, rehabilitation, and disease prevention.

Fieldwork for Healthcare: Guidance for Investigating Human Factors in Computing Systems
Editors: Dominic Furniss, Rebecca Randell, Aisling Ann O'Kane, Svetlena Taneva, Helena Mentis, and Ann Blandford

ISBN: 978-3-031-00469-8   print
ISBN: 978-3-031-01597-7   ebook

DOI 10.1007/978-3-031-01597-7

A Publication in the Springer series
SYNTHESIS LECTURES ON ASSISTIVE, REHABILITATIVE, AND HEALTH-PRESERVING TECHNOLOGIES #7
Series Editor: Ronald M. Baecker, University of Toronto

Series ISSN 2162-7258 Print   2162-7266 Electronic

Fieldwork for Healthcare:

Guidance for Investigating Human Factors in Computing Systems

Dominic Furniss

UCL Interaction Centre, University College London

Rebecca Randell

School of Healthcare, University of Leeds

Aisling Ann O'Kane

UCL Interaction Centre, University College London

Svetlena Taneva

Healthcare Human Factors, University Health Network, Toronto

Helena Mentis

Department of Information Systems, University of Maryland, Baltimore County

Ann Blandford

UCL Interaction Centre, University College London

SYNTHESIS LECTURES ON ASSISTIVE, REHABILITATIVE, AND HEALTH-PRESERVING TECHNOLOGIES # 7

ABSTRACT

Conducting fieldwork for investigating technology use in healthcare is a challenging undertaking, and yet there is little in the way of community support and guidance for conducting these studies. There is a need for better knowledge sharing and resources to facilitate learning.

This is the second of two volumes designed as a collective graduate guidebook for conducting fieldwork in healthcare. This volume brings together thematic chapters that draw out issues and lessons learned from practical experience. Researchers who have first-hand experience of conducting healthcare fieldwork collaborated to write these chapters. This volume contains insights, tips, and tricks from studies in clinical and non-clinical environments, from hospital to home.

This volume starts with an introduction to the ethics and governance procedures a researcher might encounter when conducting fieldwork in this sensitive study area. Subsequent chapters address specific aspects of conducting situated healthcare research. Chapters on readying the researcher and relationships in the medical domain break down some of the complex social aspects of this type of research. They are followed by chapters on the practicalities of collecting data and implementing interventions, which focus on domain-specific issues that may arise. Finally, we close the volume by discussing the management of impact in healthcare fieldwork.

The guidance contained in these chapters enables new researchers to form their project plans and also their contingency plans in this complex and challenging domain. For more experienced researchers, it offers advice and support through familiar stories and experiences. For supervisors and teachers, it offers a source of reference and debate. Together with the first volume, Fieldwork for Healthcare: Case Studies Investigating Human Factors in Computing systems, these books provide a substantive resource on how to conduct fieldwork in healthcare.

KEYWORDS

fieldwork, healthcare, ethnography, medical devices, HCI, human-computer interaction, health, methodology, guidance

Contents

4   Practicalities of Data Collection in Healthcare Fieldwork 57

Katherine Sellen, Aleksandra Sarcevic, Yunan Chen, Rebecca Randell, Xiaomu Zhou,
Deborah Chan, and Atish Rajkomar

5   Healthcare Intervention Studies "In the Wild" 73

Mads Frost, Cecily Morrison, Daniel Wolstenholme, and Andy Dearden

Helena Mentis, Svetlena Taneva, Ann Blandford, Dominic Furniss,
Raj Ratwani, Rebecca Randell, and Anjum Chagpar

Figures

Preface

Dominic Furniss, Rebecca Randell, Aisling Ann O'Kane, Svetlena Taneva,
Helena Mentis, and Ann Blandford (Editors)

INTRODUCTION

This is the second of two volumes designed to make conducting fieldwork in healthcare more accessible for researchers investigating human factors and technology use. Whereas the first volume reported the experiences of researchers in 12 case studies, this volume presents guidance and advice that has been developed through the collaborative efforts of a wide range of experienced researchers in the area.

The motivation for writing these volumes comes from a community need and a gap in the literature: although results from Human-Computer Interaction (HCI), human factors, and related research are reported in research publications, there is often little detail and reflection on how to conduct these sorts of studies in practice. There are exceptions within healthcare literature, such as conducting ethnographic studies in nursing research (e.g., Pellatt, 2003; Borbasi et al., 2005; de Melo et al., 2014) and guidance on qualitative research for professionals within healthcare (Holloway and Wheeler, 2013); however there is little for those who study healthcare-related technology use and work practices who come from outside healthcare.

Doing field research in healthcare is different from doing field research in many other contexts. Researchers must be sensitive to patients' experiences around their health. To protect patients and healthcare data there are often formal requirements for research governance and ethical principles that are special to healthcare, e.g., which research ethics committee within healthcare will review. There are different emotional demands in many areas of healthcare, whether in observing invasive medical procedures or talking to vulnerable people who suffer from chronic conditions, which can challenge researchers beyond normal work practices. There are challenges to gaining access and having an impact as organisational, social, financial, and administrative complexities associated with healthcare systems present barriers to progress. Also, specific advice can be given on conducting field research methods within different healthcare contexts, e.g., strategies for how to interview extremely busy nurses on the ward. Collectively these features help define why field research in healthcare is different, and why a guidebook is necessary.

These volumes are aimed at fieldwork researchers who are studying healthcare-related technology use or work practices (e.g., to improve patient safety). Both books are intended as a resource for those interested in learning, and teaching, how to conduct fieldwork in healthcare. They can inform new researchers' project plans and contingency plans in this complex and challenging domain. For more experienced researchers, this volume gives a perspective on major themes that have been developed collaboratively from researchers with different experiences. For lecturers and supervisors, it offers a source of reference and debate.

BACKGROUND

This volume represents the collaborative work that came out of an ACM Conference on Human Factors in Computing Systems (CHI) workshop, "HCI Fieldwork in Healthcare—Creating a Guidebook," which was held on 27th April 2013. During that day, 21 people from 17 different institutions discussed their experiences and case studies (a subset of which are published in Volume 1), brainstormed important topics and themes that they wanted to include in this guidebook, grouped these issues into themes that became the four middle chapters, and divided into groups to author those chapters after the workshop. Further details about the set-up of the workshop can be found in the first volume.

After the workshop, there were multiple rounds of writing and review where the editors oversaw the content, scope, and quality of each chapter and the book as a whole. As part of reflecting on the scope of the four initial chapters at the workshop, the editors decided to add two additional chapters on ethics and impact, at the front and the back of the book. While the book cannot cover everything, it was felt that these topics were deserving of their own chapters and suitable to top and tail the guidebook.

It has been a rewarding challenge to bring these chapters together into a coherent whole. They represent the experiences of many people from our international research community, and the diversity of their perspectives brings benefit to both volumes.

Even though this guidance draws on experience from an HCI perspective, we perceive these issues as having broader relevance to people in neighbouring disciplines, e.g., ergonomics, informatics, psychology, biomedical engineering, and CSCW (Computer-Supported Cooperative Work). Consequently, we preferred "human factors in computing systems" as a broader title to capture the focus of the material but maintain the breadth of issues we have covered for neighbouring disciplines. This also firmly keeps our roots in the workshop at CHI.

We take a broad view of fieldwork and healthcare. Fieldwork refers to those techniques that require the researcher to gather data "in the wild," which contrasts with surveys and laboratory studies. Fieldwork commonly involves some form of observation or interview in context. Our view of healthcare is also broad and includes physical health, mental health, emergency treatment, preventative medicine, care for chronic conditions, clinical contexts like hospitals, and non-clinical contexts like home healthcare.

Before we go further, it is worth warning readers to be mindful of extracting absolute lessons from this guidance. One of the challenges that we have faced in putting this volume together is the diversity of experiences that people have offered. Sometimes these differences have their roots in international variation in policy and practice, sometimes in variation between healthcare contexts, and sometimes just because there is an absence of hard and fast rules so expectations and advice can vary depending on who you ask. As we move forward, particularly as rules on ethics and access change, readers should make themselves familiar with current best practices for their project.

OVERVIEW: A TOUR OF THE STYLE, STRUCTURE, AND CONTENT

Style

This is pitched as a guidebook for graduates. Although it is likely to have broader appeal, it is particularly for those researchers with expertise in HCI, human factors, ergonomics, informatics, etc., who are interested in working in healthcare. Each chapter covers key themes and subthemes to be mindful of when planning and conducting fieldwork in healthcare. Readers are introduced to the material, authors share their experiences, case studies are cross-referenced from the first volume where appropriate, tips and tricks are provided, and guidance is given.

Structure

The book is divided into six chapters. The chapters progress in a roughly chronological manner, from thinking about ethical issues, to preparing for the context and networking, to developing a data collection plan, implementing a technology or practice, and thinking about impact. However, in practice these elements are inherently non-linear and can be addressed in parallel. For example, thinking about potential impacts could be one of the first things to engage with.

Content

Each chapter provides an overview of a key area to consider when planning and conducting fieldwork in healthcare.

Chapter 1 is on ethics, governance, and Patient and Public Involvement (PPI). Gaining ethical approval and access to do research in healthcare contexts is often seen as the marker that differentiates this domain from others. We provide a background for why these systems are in place, describe the main features and variances for approval, and give experience and advice on practical issues. We highlight PPI as an emerging and growing theme in the U.K., which has the potential to make healthcare research more relevant, efficient, and effective for patients and the public.

Chapter 2 focuses on readying the researcher for doing fieldwork in healthcare. It covers three broad areas: reflection on your researcher identity, handling emotions in fieldwork, and staying safe when doing fieldwork. These themes invite you to reflect on how your preferences and experiences could suit and impact different project work, as well as preparing you for the work that lies ahead.

Chapter 3 explores establishing and maintaining relationships in healthcare. You will be reliant on people for help and advice, to facilitate access to different contexts and groups, to facilitate data gathering, and to take project results forward for impact. This chapter explores how to gain access to the field and how to establish and manage relationships with clinicians and patients in different contexts.

Chapter 4 focuses on the practicalities of data collection in healthcare fieldwork. The chapter opens by contrasting how research data and contributions are different in healthcare and HCI. It then gives an overview of different methods and how these can be used to explore people's ideas and perceptions and what they actually do. Detailed and serendipitous sources of data are also covered. The chapter ends on issues of sampling and triangulation.

Chapter 5 focuses on intervention studies, i.e., those studies that design, introduce, and evaluate a change in healthcare, like the introduction of new technology. The chapter covers study design, behaving ethically and navigating governance procedures, and the practicalities of building and deploying technology for fieldwork in healthcare. An important theme here is handling the iterative nature of formative design and how this interacts with procedures for ethical approval.

Finally, Chapter 6 explores the topic of impact. The chapter takes a broad view of impact that includes impact on researchers, research, practice, and society. Experiences from different projects make these forms of impact come to life, and external resources are referred to for further information. Advice is given on how to better realise different forms of impact and how to assess them.

Overall, this volume presents an overview of the key considerations for planning and conducting HCI and human factors fieldwork in healthcare. Our aim has been to make this research area more accessible and to give graduate researchers a flying start to their projects. The companion volume, *Fieldwork for Healthcare: Case Studies Investigating Human Factors in Computer Systems*, shares concrete experiences of conducting fieldwork in healthcare—including the challenges, emotions, social dynamics, practical lessons, achievements, and disappointments experienced by researchers in this area. These are complementary volumes.

Acknowledgements

Like the first volume, this volume has been a pleasure to guide and produce. Unlike the first volume, this entire volume has been shaped and constructed by the collaborative efforts of the workshop participants. It was the conversations and ideas we shared on that day that have found their way to written text, which will form a lasting contribution. We hope researchers will be informed and inspired by this work, that more experienced researchers will be able to find support and advice, and that teachers will be able to use this material for instruction and debate.

Many people from across geographical boundaries contributed to these two volumes. We would like to thank all who contributed, both directly through editing and writing and indirectly through support in making this project happen. We would particularly like to thank the committee that supported our workshop proposal and helped us review its submissions. We would like to thank all of our workshop participants for making this project a success. Thanks to Stefan Carmien, Avi Parush, and Shari Trewin for their comments on an earlier draft of this volume. We would also like to thank Ron Baecker and Diane Cerra for their support and advice.

WORKSHOP COMMITTEE

Gregory Abowd, Jonathan Back, Ken Catchpole, Gavin Doherty, Geraldine Fitzpatrick, Ioanna (Jo) Iacovides, Josette Jones, Tom Owen, Madhu Reddy, Mark Rouncefield, Penelope Sanderson, Chris Vincent, Robert Wears, and Stephanie Wilson.

WORKSHOP PARTICIPANTS

Ann Blandford, Anjum Chagpar, Deborah Chan, Yunan Chen, Andy Dearden, Mads Frost, Dominic Furniss, Kristina Groth, Brian Hilligoss, Cecily Morrison, Aisling O'Kane, Rebecca Randell, Atish Rajkomar, Raj Ratwani, Aleksandra Sarcevic, Katherine Sellen, Svetlena Taneva, Anja Thieme, Ross Thomson, Heather Underwood, and Xiaomu Zhou.

CONTRIBUTIONS TO SPECIFIC CHAPTERS

Thanks to www.becandderek.co.uk whose training materials informed some of the PPI advice we have included in Chapter 1.

ILLUSTRATIONS

Thanks to Smaragda Magou who drew the illustrations. Furniss brainstormed the ideas after the chapters were completed, which Smaragda interpreted and transformed into graphic form.

FUNDING

Furniss, Blandford, O'Kane, and Rajkomar were funded by the CHI+MED project, supported by the U.K. Engineering and Physical Sciences Research Council [EP/G059063/1].

Thomson acknowledges support from the EPSRC through the MATCH programme [EP/F063822/1].

Dearden and Wolstenholme's contribution represents independent research by the Collaborations for Leadership in Applied Health Research and Care for South Yorkshire (CLAHRC SY). CLAHRC SY acknowledges funding from the National Institute for Health Research (NIHR).

Hilligoss's research was funded in part by a health services research dissertation grant (number R36HS018758) from the U.S. Agency for Healthcare Research and Quality.

The views and opinions expressed in this volume are those of the authors, and not necessarily those of the U.K. NHS, the U.K. NIHR, the U.K. Department of Health, the U.S. Agency for Healthcare Research and Quality, or any other affiliated body.

CHAPTER 1

Ethics, Governance, and Patient and Public Involvement in Healthcare

Dominic Furniss, Rebecca Randell, Svetlena Taneva, Helena Mentis,
Daniel Wolstenholme, Andy Dearden, Aisling Ann O'Kane, and
Ann Blandford

This chapter introduces you to elements of research design and conduct that often distinguish research in healthcare from other contexts. Researchers often bemoan the laborious ethical and governance procedures involved in doing research in healthcare. However, if we can understand the intention and concerns of ethics review boards and appreciate the advantages of ethics review it may be perceived in a more positive light, be less onerous to work with, and make research in healthcare more accessible to practitioners and researchers.

This chapter covers three broad areas focused on important considerations for the design and conduct of your research in healthcare—ethics, governance, and patient and public involvement (PPI). These elements have a strong overlap too. For example, involving patients and the public in the design of the study can inform ethical decisions and governance details such as the wording of lay summaries and information sheets for participants. Ethical principles and concerns will impact the governance of the research, and vice versa. We highlight the content of this chapter in the table below:

What do I need to know about research ethics in healthcare?	This section provides an introduction to the historical context of ethics in healthcare and an overview of governance procedures and advice needed for studies in healthcare.
How do I get research ethics approval?	This section highlights the benefits of ethical review to researchers and indicates the sort of timeframe needed for this process. We reflect on the thorny subject of whether studies always need full ethical review and provide lessons learned.
How could Patient and Public Involvement (PPI) help me?	Involvement concerns itself with getting advice and guidance on research design, analysis, and dissemination. Involvement is important for designing relevant studies in the right way, considering the perspectives of patients and the public. We include healthcare practitioners as a group that should also be involved.

We do not intend to repeat or replace the many informative online resources about these topics. Indeed, the reader would do well to seek these out for the latest advice in what is a changing landscape, and to understand local practices. Instead we want to convey what is important and what we have learned from conducting fieldwork in healthcare, from a human factors and HCI perspective. We should emphasise that this chapter has a U.K. slant because that is where the lead authors are based. However, our co-authors from Canada and the U.S. have helped highlight similarities and differences with their systems. Both terminology and some of the details of processes and practices vary across healthcare systems, and you should familiarise yourself with local practices. Many of the tips and the guidance we give apply equally well across borders.

1.1   WHAT DO I NEED TO KNOW ABOUT RESEARCH ETHICS IN HEALTHCARE?

Research ethics can be a source of considerable challenge for HCI fieldwork in healthcare settings—it can bring both benefit and frustration to the project. If your study involves human subjects as participants in research, then you will need some form of ethical approval. If your research directly involves healthcare services or service users then it is likely to require ethical approval from a committee or board within the healthcare system. These are known as Research Ethics Committees (RECs) in the U.K.; Institutional Review Boards (IRBs) in the U.S. and Canada; and also Research Ethics Boards (REBs) in Canada. However, as we shall see later in the chapter, there are instances where ethical approval outside of the healthcare system is sufficient, e.g., via a university ethics committee.

We distinguish two interrelated levels of ethics: the higher level of ethical principles and the more detailed level of governance. In our experience, the main challenge comes not from any difficulty that HCI or interaction design researchers face in behaving ethically (at the higher level of ethical principle). Instead, a greater challenge can come from the complexities that arise in navigating the specific processes for the ethical governance of research in healthcare (at the detailed level of process). These two are linked, as it is this detailed level of study design that determines how principles are respected and delivered, so attending to the detail of ethical governance is essential. In the following subsections we first outline the background to ethical principles in healthcare research. We then look at the details of ethical governance in healthcare to make the standards and procedures, and why we follow them, more transparent.

1.1.1   THE HISTORICAL CONTEXTS THAT DRIVE HEALTHCARE'S ETHICAL PRINCIPLES

Every discipline carries its own guidelines for ethical behaviour which are shaped by the particular history, practices, and concerns of that discipline. Healthcare professions usually trace the history of their ethical thinking to the 5th century BC and the Hippocratic Oath. However, for research ethics, the more relevant document might be the Nuremberg Code of 1947 (HHS, n.d.) which is a set of ten principles that form a code of conduct for conducting ethical research with human subjects. This document stipulates that the voluntary consent of the human subject is absolutely essential. The statement came from the judges' verdict at the Nuremberg Trials following attrocities conducted by Nazi scientists during World War II. Research is rightly subject to careful ethical governance given some historical cases: the experiments conducted by Nazi doctors and scientists on concentration camp prisoners during World War II; the Tuskegee Study of Untreated Syphilis in the Negro Male (CDC, n.d.) conducted by the U.S. Public Health Service (1932–1972) where medical researchers intentionally withheld treatment from African-American patients with syphilis; issues in conducting medical research in developing countries such as situations where drugs might be trialled with populations in developing countries, but the typical participants in the trial would never be able to afford the drugs even if they were proved to be successful (Bhutta, 2002); and, in the U.K., the scandal involving the unauthorised removal, retention, and disposal of organs at Alder Hey children's hospital (Redfern et al., 2001).

Subsequent to the Nuremberg ruling, in 1964 the World Medical Association (also initially established in 1947 following World War II) adopted the "Helsinki Declaration," with subsequent modifications (World Medical Association, 2008). This declaration provides much of the basis for ethical governance of medical research internationally, being interpreted by researchers in different countries (e.g., Emmanuel et al., 2000). Its principles include respect for the individual, their right to self-determination and to make informed decisions, and the need for special ethical vigilance when involving vulnerable groups.

The basic principles of the Helsinki Declaration should not present any particular difficulty for HCI and design researchers. However, the way in which different countries interpret, formalise, and enact the principles in specific legislation and practical governance procedures could challenge "business as usual" for HCI and design researchers. In our experience, it is at this level of very detailed procedure that miscommunication and misunderstanding arise between HCI researchers and ethical governance bodies in healthcare. In other domains research governance procedures are not as extensive, detailed, complicated, or strict.

1.1.2   RESEARCH GOVERNANCE IN HEALTHCARE

The word "governance" is derived from "governing," and simply refers to the manner in which things are managed. Ethical governance includes the detailed design and management of a research study, which is essential to how ethical principles are enacted (as explained above).

In the U.K., the Research Governance Framework (Department of Health, 2005) provides a comprehensive summary of research governance in healthcare. It says: "The public has a right to expect high scientific, ethical and financial standards, transparent decision making processes, clear allocation of responsibilities and robust monitoring arrangements." The Research Governance Framework describes:

- arrangements to define and communicate principles, requirements, and standards;

- delivery mechanisms to ensure that these are met; and

- arrangements to monitor quality and assess adherence to standards nationally.

Five domains are covered in the document: ethics; science; information; healthcare, safety, and employment; and finance and intellectual property. Key principles are as follows.

- The dignity, rights, safety, and well-being of participants must be the primary consideration in any research study.

- Informed consent is at the heart of ethical research (this topic is discussed in detail in Chapter 3).

- Research in healthcare cannot be conducted until it has received a favourable opinion by a Research Ethics Committee (REC).

- Peer review is essential to ensure the scientific quality of research. Research that does not usefully contribute to knowledge is unethical. Researchers with the right skill sets must conduct the work to high standards.

- All research staff should know their legal and ethical duties to ensure the appropriate use and protection of patient data. Particular attention should be given to systems that ensure the confidentiality of personal data.

- Data should be retained for an appropriate period so analysis can be done after the study, and appropriate documentation must be kept so the project can be monitored and audited.

- There should be free access to information on the research being conducted and on the findings of research. Efforts should be made to making this accessible, e.g., by taking language into account.

Similar standards and advice are given in the U.S. and Canada.

In Canada the main documents to refer to include the Tri-Council Policy Statement: Ethical Conduct for Research Involving Humans (TCPS-2), which details the regulation and requirements around performing research in healthcare. For example, it governs general healthcare research requirements for work with human subjects, as well as country-specific considerations such as doing research in multiple jurisdictions and working with First Nations individuals. This policy is applied in conjuction with the Personal Health Information Protection Act (PHIPA) 2004, which regulates and protects the collection, use, and disclosure of patients' personal health information.

In the U.S., the Belmont Report is the foundational document of the current system of U.S. human subjects protections. It outlines three key ethical principles for conducting research with human subjects: respect for persons, beneficence, and justice. The Belmont Report, in turn, informed the U.S. Department of Health and Human Services' (HHS) Code of Federal Regulations (45 CFR 46) in 1974, which, among other things, defines what constitutes "research" (45 CFR 46.102(d)) and what constitues a "human subject" (45 CFR 46.102(f))[1]. In 1991, Subpart A of these regulations (Basic HHS Policy for the Protection of Human Research Subjects) was adopted by 15 federal agencies and became known as "the Common Rule."

In addition, privacy protections for health information are required by the U.S. federal law Health Insurance Portability and Accountability Act of 1996 (HIPAA). HIPAA's protections focus on "individually identifiable health information," which HIPAA defines as information in "any form or medium" that "[r]elates to the past, present, or future physical or mental health or condition of an individual; the provision of health care to an individual; or the past, present, or future payment for the provision of health care to an individual." (See definitions in 45 CFR Parts 160, 162, and 164.) This is also referred to as protected health information (PHI). HIPAA's privacy provisions apply to protected health information in "any form or medium." That means everything containing PHI: paper records as well as electronic ones, faxes, emails, exchanges in telephone conversations, and even just talking face-to-face. If it is health data, and it is identifiable, it is covered.

For researchers, the key bits to know are that HIPAA privacy protections supplement those of other federal regulations (e.g., the Common Rule), HIPAA protects protected health information (PHI) originating or held in covered entities such as health providers (see online guidance[2] to determine what constitutes a covered entity), and research activity using PHI generally requires authorisation.

Researchers from other countries should refer to their local practices. Where this is absent, established forms of advice like those mentioned above can be used to promote best practice.

1.2 HOW DO I GET RESEARCH ETHICS APPROVAL?

Review and oversight from research ethics review boards and committees is a fundamental principle for being allowed to conduct research in healthcare. This can bring with it benefit and frustration. Those who have a more positive stance toward research ethics review see it as an intergral part of the design and process of their research, and part of quality control for research in healthcare. Those

---

[1] http://www.hhs.gov/ohrp/humansubjects/guidance/45cfr46.html#46.102.
[2] http://www.cms.gov/Regulations-and-Guidance/HIPAA-Administrative-Simplification/HIPAAGenInfo/Downloads/CoveredEntitycharts.pdf.

who have a more negative stance toward research ethics review see it as a laborius, unaccommodating, bureaucratic process that delays research disproportonately to the risk associated with many HCI and human factors studies. Your experience could include a mix of these. The more you can embrace the positive stance, the more satisfying and rewarding this stage of the research will be.

Tip: In some countries, including the U.K., proportionate and expedited review processes within the healthcare system have been introduced for studies considered to have no material ethical issues (discussed further in Section 1.2.3).

1.2.1   THE BENEFITS OF RESEARCH ETHICS REVIEW

Preparing for healthcare research ethics review involves filling out comprehensive documents that force you to confront the specific details of how you will conduct the study and provide clear justification for your decisions. For example, on recruitment you might need to report: who you will recruit, what inclusion and exclusion criteria there are, how many people, how the sample size was determined, who will approach potential participants and how, and how consent will be sought. Each part of your project will need to be thought about thoroughly, which helps you plan the study.

Having your research plans independently reviewed has a number of benefits. Firstly, the review board or committee can highlight ethical issues, details of clinical contexts, and patient sensitivities that you are unfamiliar with and may not have considered. Secondly, this stringent review can give credence to a study, giving you and others more confidence in its feasibility and value. Thirdly, having research ethics approval should reduce the likelihood of something going wrong, because of the scrutiny of the study's procedures. Finally, some journals will not accept submissions that have not gone through appropriate ethical review procedures.

1.2.2   FRUSTRATIONS CAUSED BY RESEARCH ETHICS REVIEW: HOW LONG WILL IT TAKE?

One of the main frustrations of going through research ethics review within the healthcare system is the time that it takes. The horror stories you hear are enough to put people off: nine months, a year, or even two years to get ethical approval. This can be a real issue and some funders in the U.K. have now taken to witholding funds until the ethical review process is completed or close to completion. However, these need to be balanced with more positive stories (e.g., approval gained within three months) and further explanation needs to be given so these differences in timescales can be understood, as discussed below.

First of all, what do these timescales include? You need to think about the time required to write all the documentation, not just the time between submitting it to the review committee and getting a positive response. Also, in the U.K., once you have approval from your NHS (National Health Service) research ethics committee, you need further approval from the local sites (e.g., individual hospitals or clinics) that are involved in the study, which can take just as long. So, you need to be aware that the timescales that people talk about can include very different things.

At one extreme: if you are very new to healthcare, it is likely that you will not know where to start, you will not know who to talk to, you will not know any of the processes and forms, and you have no precedent to build upon, e.g., previous information sheets and consent forms that can be adapted. In these circumstances, six months to a year or more is not unreasonable for all approvals. The good news is that you have this book, which is intended to give you a good start, and there are many supporting materials and sources of advice online to help you.

> **Tip:** If you are new to healthcare, try to partner with someone who is more experienced, who can open up relevant networks and doors, and provide access to similar documentation, to facilitate your work. Key study enablers who are stakeholders in the research and work in healthcare (i.e., on the inside) can greatly facilitate access—described further in Chapter 3.

An example of ours which took a long time is reported by Furniss (Volume 1, Chapter 3). The research was looking at medical device usability and errors in an oncology outpatient unit. Starting from scratch to build a research network for support and access (including meetings with senior hospital staff) and to develop protocols took about three or four months. The NHS REC took about two months to review the proposal after we had submitted it to them. We only had one minor issue: they wanted some information on our consent form to be transferred to the information sheet. However, frustratingly, it took them two months to send a letter to inform us of the detail of this request. We made the changes in about a day and sent them back for approval, which took another month. All in all, with some support from a clinician on our project who was able to make introductions, the whole process took about nine months from start to finish. It must be noted that, even at the time of writing, the NHS Health Research Authority (HRA) in the U.K. is implementing new processes to address these criticisms.[3] This is both an indication of ongoing progress, and of the constantly changing landscape of research governance and approval which emphasises the need to check the latest rules and regulations for different projects.

At the other end, if you are experienced with doing research in healthcare, you have a detailed grant proposal which can be adapted to suit the required forms needed for ethical review, you already have a supportive network, you have been through ethical review before, then three months is more reasonable. You can complete the required forms in a week or two. You can get ethical clearance within two months if everything goes well. You can get local site approval if required in another month or two (e.g., this extra approval is often needed in the U.K.). From this more

---

[3] http://www.hra.nhs.uk/research-community/booking-submission-changes-spring-2014/.

advanced perspective you are developing the detail of the project, completing the administrative steps, and finalising access.

Figure 1.1: Access and Approval Snakes and Ladders. Gaining ethical approval and site access can seem like a game of snakes and ladders. A bit of luck is needed but it's good to be aware of the pitfalls and the positive strategies that can affect progression. What will your game look like?

An example of ours which took an exceptionally short time was a study conducted by Healthcare Human Factors, the company that Taneva works for in Canada. The research was looking at the experience of kidney disease patients and their providers, focusing on their information and communication needs. The research team introduced the project to the department leads at a Canadian hospital via teleconference and emailed them the proposed study protocols. The department leadership were keen to take part in the study and wanted the research team to begin the study the following week. The research team emailed documents through to the department for review on Friday, the department's questions were received and addressed by the team Monday morning, on Monday afternoon the project proposal was submitted to their IRB, and approval was granted within four hours. Here the majority of the protocol documentation was prepared, senior leadership were very keen on the study and had notified their IRB about the urgency of this upcoming proposal, it wasa low-risk study, there was pressure to expedite the study, and the research was being conducted by a very experienced team at a large teaching hospital. This is an exceptional case and would not always be possible, especially in countries where review processes take a longer timeframe.

1.2.3   DOES MY STUDY NEED FULL ETHICAL REVIEW AND APPROVAL?

Review boards and committees have been set up to review ethics and governance procedures for projects defined as research taking place in healthcare. The Health Research Authority, in the U.K., states on their website: "For the purposes of research governance, 'research' means the attempt to derive generalisable new knowledge by addressing clearly defined questions with systematic and rigorous methods." In the U.S., the Department of Health and Human Services defines research as "a systematic investigation including research development, testing and evaluation, designed to develop or contribute to generalizable knowledge".[4] The simple answer is that all HCI and human factors research involving human subjects will need some form of independent ethical review. If this involves health services and their users then it is likely to require review within the healthcare system, i.e., via a REC, IRB, or REB.

A full ethical review within the healthcare system will involve all the relevant documents being reviewed when a committee sits, and the researcher may also be required to attend for clarification and questioning. It is worth noting that the vast majority of qualitative studies that are considered low-risk with no material ethical issues will qualify for an *expedited* (U.S./Canada) or *proportionate* (U.K.) review process. You should seek local guidance to see if your study qualifies for this route; e.g., in the U.K. a "No Material Ethical Issues Tool" can be found online. A proportionate review can involve documentation being reviewed by a few committee members by email. This saves time at the review stage but the same protocols and forms still need to be completed.

There are some instances where your study might not need approval through the health service ethics processes, but might be reviewed by other processes. In the U.K., studies that are exempt from NHS REC review typically use university ethical clearance processes. NHS REC review is not normally required for research involving NHS or social care staff recruited as research participants by virtue of their professional role.[5] Also, studies of technology use outside formal healthcare settings, where participants are recruited as citizens rather than patients, would not nor-

[4]   http://www.hhs.gov/ohrp/humansubjects/guidance/45cfr46.html#46.102.
[5]   http://www.nres.nhs.uk/EasySiteWeb/GatewayLink.aspx?alId=134016.

mally require NHS REC review. For example, Attfield et al. (2006) conducted a study of people's health information seeking, recruited from the general population, and as such they obtained ethics approval through the university's ethics processes.

Research projects (as defined by the NHS REC) are not the only type of projects to take place in health. Other recognised types of project include service evaluations and audits (there are many different terms for such studies in the U.K. and across the globe). Service improvement and audit projects tend to be undertaken with a view to improve local services and, in the U.K., are governed by the *Clinical Governance* Framework. *Service improvements*, *service audits*, and studies that operate at a *pre-protocol* level (i.e., doing preparatory work for a quality improvement intervention or design before a larger "research" study), are discussed further in Chapter 5, this volume.

> **Tip:** It is always a good idea to seek advice from the health organisation's research or clinical governance department, or the NHS REC or equivalent for their advice, before putting an application togther. This is particularly important where there is uncertainty around the "type" of project being undertaken.

According to the definitions of research above, the defining feature of research studies is their intent to produce generalizable knowledge. As a heuristic, some consider research to be a study with the intent to publish. However, research studies may not always intend to publish, and service improvements and audit studies can also be published. For the latter cases you should check the author guidelines of the journals you wish to publish in to see what their requirements are. It is likely that empirical research involving patients would require a formal ethics approval, but for a project classed as something else (e.g., a service evaluation or an audit) a different form of approval would be sought. In any case, you need to follow a recognised path to get the appropriate approvals for your research project. Depending on your project this could be through gaining approvals at the university or in the healthcare system. Seek advice to determine which one is right for your circumstances.

An understanding of what does and does not require full NHS REC review might help guide a student to the type of project to undertake if time is a real constraint. For example, in the U.K., Randell normally advises MSc students to select projects that only involve interviewing staff as this does not currently need ethical review from healthcare bodies. In these cases, university ethical approval is adequate, although you will still need local research *governance* approval at the site you use for recruitment. Ideally, you would be able to plan ahead to get the ethical approvals you need in time for the project you wish to do. For supervisors, this could mean starting appropriate ethical review channels before a student has been recruited to conduct the project.

A final note around process is that regardless of how your project is classified if you are stepping foot into a health institution, you will need some sort of organisational approval. This tends to be in the form of a research passport, honorary contract, or letter of access. The requirement stems from you signing something to say that you will abide by the confidentiality required by the organsiation and that they know you are on site so that you are insured if anything goes wrong.

In the U.K. where your project is classed as research by the NHS REC, and you will work across a number of sites, you can apply to the "host" organisation to prepare a "Research Passport" for you. This is a document that shows to any other sites at which you are undertaking research that the checks (Criminal Record Bureau, Occupational Health, agreement to confidentiality policy, etc.) have been undertaken and evidenced. This stops duplication of these checks at each site. Accessing NHS property for research purposes could also involve Human Resources producing a Letter of Access or an Honorary contract. There might not be a uniform process for this so it is best to follow local advice. Having the piece of paper that says that you have all the necessary permissions to be in the hospital is a good thing to have and can be empowering for the researchers; it also stops any unpleasant situations where you are invited in by the well-intentioned, enthusiastic clinician, without anyone in "management" knowing that you are working in the area.

The main take-home message for this section is that you should seek official guidance from established researchers and research supervisors, and speak to people involved in coordinating the work of IRBs and RECs to get their advice on what approval your study needs, and to get realistic completion time estimates for these processes. However, bear in mind that this can be a grey area, and different people, committees, and boards might have different opinions about what the correct thing to do is. Other forms of guidance are given below.

> **Tip:** When seeking advice, start with the published guidance online in your relevant country and then talk to colleagues who have done similar work before, to find out who they spoke to and the routes they took. University ethics committees and research and development departments within hospitals might also be good places to start.

1.2.4   EXPERIENCES AND LESSONS LEARNED IN HANDLING RESEARCH ETHICS APPROVAL IN HEALTHCARE

This section outlines experiences of managing the ethics and governance procedures on the frontline and lessons we have learned along the way.

Start Talking and Start Writing Early

Starting the ethical review process can be daunting. Things that make it such include the myriad of different rules, procedures and processes, and the extensive documentation that needs to be completed. Start talking to people early on—one of the best sources of advice will be experienced researchers, as well as the institution's research and development department, or the review board/committee's coordinator. These people should be able to guide you or know where you can find answers. Also, once you know what documentation and requirements are needed, start putting together your proposal. Writing forces you to commit to protocols and think things through thoroughly. This process of writing and reflection may raise more issues that you need to seek advice on.

Variations in Ethical Review Assessments

If navigating complex ethical issues and governance procedures for healthcare research was not hard enough, then variance in this process can add further uncertainty and surprises into the mix. This variance can start before formal ethical review—in terms of deciding which are the appropriate bodies to provide ethical review of your study. It can also happen during the project if different sites have different demands and requests. Additionally, there can be controversies after studies have been completed—for example, there are some high profile cases of quality improvement studies that arguably should have sought ethical review (Thompson et al., 2012). If in doubt, the best thing to do is ask; a REC or similar point of contact can give you a formal statement of exemption.

Variations related to ethical review are by no means the norm, but you should be aware of them. Thompson et al. (2012) report on a multisite study in the U.S. and describe the different requirements and timescales of different IRBs they engaged with. In our experience, variations can sometimes be for specific practical and ethical reasons, different processes and procedures, and sometimes they might be based on differences of opinion. This highlights the importance of seeking local expertise and advice.

Be Firm, but Not Uncompromising, where a Certain Methodology Needs to be Followed

With the benefit of hindsight and the added confidence that experience brings, we believe you should be clear and assertive about feasible and fitting research processes. Ethics committees want to make sure that good research gets done, so be clear about the value of the research, why you want to use particular data collection methods, and the consequences of those for the robustness of the findings. If, for example, video recording is the only way that you will gather the necessary detail, make that argument clearly. At the same time, you need to show that you have thought about the consequences of your data collection methods for the people in the setting, what their concerns about your methods may be, and what you will do to address and minimise those concerns. In planning your methods of data collection and obtaining consent, talk to those in the setting about what they consider to be feasible and get advice from PPI groups about how and when to approach patients for consent. Because of differences between ethics committees, in the U.K. at least, it is worth talking to other researchers in your institution about their experiences with different ethics committees.

Tip: Make clear and firm arguments why certain methodologies are most appropriate. You should state how each method will impact the quality of data and propose mitigation strategies to alleviate ethical concerns.

The dilemma is that sometimes we only learn what is feasible and fitting with experience. In this case, an experienced advisor to the project can be a good source of support. You can get advice from healthcare professionals regarding what is feasible, and it is good to describe what consultations you have already conducted within your ethics application. PPI can be used as a form of consultation with patients and the public, which is discussed later in the chapter. Where possible, some form of pilot study could assist the learning process, e.g., Chapter 5 discusses pre-protocol research which can test the feasibility for a larger study. The case studies in Volume 1 also provide examples of what is feasible, in some cases providing detail on how the ethics approval process was managed (e.g., Volume 1: Chapter 2).

Be Clear about What Is Practical

We have been in positions where our plans for consent have been well intended, approved by clinicians and the REC, but they have not proved feasible or fitting to practice. For example, in some settings we have found that staff members do not have time to each sit down with the researcher for 15 minutes to go through the planned informed consent process. Instead we were given three or four minutes at the start of a shift to describe to the whole team who we were and the main points of the study. When wanting to observe a two-minute interaction of a nurse setting up an infusion pump on a patient, it is disproportionately disruptive to their work to stop everything and spend 15 minutes gaining consent from the patient when they are not the focus of the study and no patient identifiable information is being recorded. Worse still is that if a patient has difficulty understanding the researcher due to their competence in the local language, their condition, or the drugs they are on, they may get very concerned about what you are asking them to sign and commit to. In this case, it is much more proportionate and fitting with the context to shadow the clinician and give a brief introduction to the patient of who you are and what you are doing, and politely ask whether they mind if you observe their treatment. If approved protocols are not workable then they should be revised and re-approved by the relevant board or committee. Randell (Volume 1: Chapter 5) describes some of the challenges of gaining consent from patients, particularly when they are in a distressing situation.

Tip: Be clear about what is practical, which includes the potential disruption to patients and healthcare work. If you find your processes are impractical when you have already started your study, you will need to return to the ethics committee to adapt the procedures.

What Should Be Included in an Information Sheet and Consent Form?

Information sheets and consent forms can be an important part of the consent procedure, to ensure critical information is conveyed in a written form, and that consent is recorded in writing.

Information sheets should typically not only describe the study's aims and methods (in non-technical language), but also:

- explain why the patient or member of staff has been asked to take part;

- make clear that participation is voluntary and that they can later withdraw from the study at any time (and explain what will happen to the data if they do withdraw);

- explain what will happen to them if they decide to take part;

- describe any possible risks or benefits of participation in the study;

- state that any data collected that relates to them will be anonymised and their personal information will be kept confidential, and explain how this will be achieved;

- explain what will happen to the results of the study (e.g., will they be used to inform the design of a technology? Will they be published in a journal? Will the participant receive a summary of the results?); and

- include contact information so that participants can contact the researchers afterwards.

Consent forms should not only capture the participant's agreement to take part but should also record: (i) that participants have read and understood the information sheet; (ii) that they have had an opportunity to consider the information, ask questions, and clarify anything they do not understand; and (iii) that they understand that their participation is voluntary and that they can withdraw at any time without giving any reason. The consent forms could also contain contact information for the research team and details of how the consent form was presented (e.g., sometimes a participant needs or prefers for it to be read to them) and by whom. The consent form can be broken down into various sections so the participant can initial (not tick) that they have read and understood each section.

Patient and Staff Posters and Flyers

In some instances getting full written informed consent from everyone present in the environment where an observational study is occurring is infeasible. For these instances, some review committees prefer that posters and flyers are printed to inform people about the study. Posters and flyers can be used in conjunction with more formal consent procedures.

Consent Waiver

Processes should push for gathering appropriate data, while ensuring that participants' personal information is not compromised. Be clear about the consequences of your approaches for gaining consent, for both the people whose consent is being sought and the researcher. For example, gaining written consent from all patients being discussed by providers during observations may be infeasible; it may require so much time that data collection is negatively impacted, and it may mean approaching patients at what is already a distressing time. In such cases, you can make an argument for not gaining written consent, but you need to be clear about how you will protect the privacy of the patients being discussed. This can be done through anonymity, which avoids the recording of personal identifiers so data cannot be linked to participants, and confidentiality which concerns who can access information—usually only the research team can access confidential data to do with a study. For example, in the U.K., Randell gained ethics approval to video record multidisciplinary team meetings without consent from the patients who would be discussed. The ethics committee accepted this, but key was explaining why it was not feasible to get consent from patients and what would be done with the data, i.e., explaining that the study was not interested in the details of individual patients so no patient identifiable data would be transcribed and that the sound from the video recordings would be deleted after a certain time. The issue of waiving consent is also addressed in this volume in Chapter 3, as well as in Volume 1: Chapter 2.

Alternative Forms of Consent

Other forms of consent are also worth exploring. Whether these are allowed will depend on the risks to those individuals taking part in the research and other circumstances. Examples of alternative forms of consent include:

- Phased consent—this allows the opportunity for the participant "to renegotiate as required throughout an emergent research process" (http://www.lancaster.ac.uk/researchethics/4-3-infcons.html)

- Passive consent—this is where staff or patients are informed about a study and if they do not explicitly dissent, then consent is assumed.

- Delayed consent—this can happen in emergency situations where gaining consent is not possible in advance of the research and where gaining consent would make the study impossible, e.g., after a cardiac arrest. Consent should be sought as soon as possible after the incident (http://www.rcn.org.uk/__data/assets/pdf_file/0010/78607/002267.pdf).

- Implied consent—this is where written and verbal consent is not given but consent is implied, e.g., when a participant returns an anonymised questionnaire, their consent can be implied (http://www.rcn.org.uk/__data/assets/pdf_file/0010/78607/002267.pdf).

- Consent by proxy—this is where participants are unable to give consent on their own, typically adults without capacity to consent, e.g., due to confusion caused by medication, or children, so consent/assent is sought from a family member or other person to include an individual in research (http://www.rcn.org.uk/__data/assets/pdf_file/0010/78607/002267.pdf).

Split and Stagger Multi-Phase Studies

One practical strategy for dealing with ethical approval is to split your study into multiple parts, and apply for ethical approval for each part separately. This approach can mean that you get ethical approval for one part and get started with that aspect of the research while waiting for ethical approval for the other parts. We recently used this approach for a study in the U.K.. The first part of the research involved telephone interviews with healthcare professionals, which currently does not require approval from an NHS REC in the U.K., so university research ethics approval and the necessary local NHS governance approvals were gained. A separate REC application was submitted for the later parts of the study that would involve video recording in the operating theatre and getting consent from patients. This approach is also described by Sarcevic (Volume 1: Chapter 2).

Amendments Can Reduce Time and Effort

Rather than splitting your study per se, you may complete a study and then plan a similar follow-on study. In this case we suggest you consider amending an IRB/REC/REB proposal that has already passed review, rather than starting from scratch. This can save considerable time as described by Furniss (Volume 1: Chapter 3).

Templates and Umbrella Approvals

If senior researchers are in a situation where they will be doing multiple studies, then they may have previous documents to adapt and build from. We have heard of some research teams getting approval for studies in general terms: this approval acts as an umbrella for different studies that fall under its remit. Both of these things could greatly facilitate the organisation of your study.

Dealing with Ethical Dilemmas

Ethical dilemmas can be expected by researchers or they can arise unexpectedly from patients, practitioners, and IRB/REC review. To prevent lengthy delays, you want to aim to foresee all the dilemmas related to your study and have solutions to mitigate them before the study goes to IRB/REC review. Example dilemmas include:

- If you are doing an observational study and observe a serious error, what do you do?

- If you interview staff on their knowledge of volumetric infusion pumps and you become greatly concerned about the lack of knowledge of a particular member, how do you balance patient safety with participant confidentiality?

- If you have video data of staff and have promised that it will remain confidential, under what circumstances would this video be shared? For example, what would happen if management demand to see it, or if the police demand to see it, or if a court orders that they see it?

- If you want to study treatment related to head injuries and concussion, arguably people will not be able to give informed consent once they are in this state. How could you gain informed consent from these people?

These issues will need to be thought through thoroughly. During the conduct of the project, depending on its size, it might be worth considering establishing a "safety committee" to review sensitive ethical issues, e.g., if intentional harm is suspected or if incompetence is a concern (WHO, 2013).

Talking to practitioners, patients, and the public to get their perspective on dealing with ethical dilemmas can be enlightening as well.

1.3   HOW COULD PATIENT AND PUBLIC INVOLVEMENT (PPI) HELP ME?

1.3.1   PARTICIPATION, ENGAGEMENT, AND INVOLVEMENT

In this section we focus on involvement, which brings in different patient communities to get their perspectives on research. Their perspectives may well be different from your own. Involvement may be understood by comparing it with two other modes of interacting with people within and around research projects—participation and engagement.

- Participation is the traditional form of interacting. This is where people actually participate in a study, sign consent forms, and contribute to study data to be analysed.

- Engagement is where you seek to interact with people who might not have heard of the study before. This might be to disseminate your findings more broadly or to raise awareness about a particular issue.

- Involvement is about getting people's input on a study's design, analysis, dissemination, research priorities, and wording of participant materials.

It is useful to reflect on all three modes of interacting with people to see if you might be missing an area that would benefit your study, but our focus in this section is on involvement.

1.3.2   INVOLVING PATIENTS, PUBLIC, AND PRACTITIONERS

As human factors and HCI researchers, conducting research in healthcare typically involves engaging with communities and settings that are unfamiliar to us. Appropriately involving people in the design of the research is important to make sure that we are solving issues that are meaningful to them; that the study is designed appropriately; that materials are worded in a simple way, without jargon; and that our dissemination plans are effective.

The importance of PPI has been recognised for healthcare research projects in the U.K. for a number of years. The U.K. National Institute for Health Research (NIHR) now requires that this be appropriately considered in projects that they fund. Similarly, the REC application form requires researchers to explain in which aspects of the study patients and public have been or will be involved and to provide justification if patients and the public will not be involved. Even though this is not a requirement in many other countries, it may still be a source of advice for studies that would benefit from involving patients and the public.

To find out more about how PPI has been used effectively, a good summary that includes its impact on different phases of research projects, and impact on the researchers, participants, and the wider community can be found in Staley (2009). Furthermore, Stewart (2011) and Faulkner et al. (2013) have both edited a collection of case studies that show positive PPI impact on recruitment for research studies, study design, outcomes, and informing research agendas across major clinical areas.

We have included healthcare practitioners as groups to involve with research activities because the same goals, principles, and methods apply. We need to involve practitioners to advise us on research priorities, design, and dissemination to ensure that our research is effective.

Figure 1.2: Ivory Tower Detachment. From the dizzy heights of the ivory tower to real need on the ward—how do we bridge this gulf to deliver meaningful impact in practice?

1.3.3   INVOLVEMENT: WHEN AND HOW?

The PPI handbook (NIHR's RDS PPI handbook, n.d.) suggests five key stages in the research process where involvement could take place.

- Design of the research: this can inform and clarify methodology and help with recruitment strategy, e.g., advising on when it might be appropriate to approach patients for consent.

- Development of the grant application: this can ensure some aspects of ethics have been considered and can also inform where PPI could be used throughout the project.

- Undertaking/management of the research: this can include assisting in designing protocols, providing a patient perspective on the acceptability of the study protocol, and assessing the readability and suitability of patient information sheets and consent forms.

- Analysis of data: this can provide a different perspective on the data, as seen from the viewpoint of a patient.

- Dissemination of research findings: this can include helping to produce patient-friendly summaries of the research findings, suggesting strategies for dissemination so that your findings reach a wider audience, and helping to present findings with the rest of the research team.

We have found involving patients and the public to be valuable for our research. A patient representative recently reviewed a research proposal that aims to investigate human error around infusion pump use. They highlighted a concern that had not come up from our research perspective: whereas we wanted to find out as much as we could about errors and inform staff and patients about the importance of these issues, the patient representative said that this could erode confidence in the services they rely upon and that sometimes "blind faith" is preferable to being fully informed of all the risks and potential deviations in care. This has led us to rethink how we engage with patients in this area. We have planned a PPI workshop at the beginning of our project to get a broader range of advice on this issue. This early workshop will also review the appropriateness of patient information sheets and consent procedures that relate to the project.

There is a range of strategies for involving patients and the public in your research. Patients and public representatives can be co-opted onto project steering committees if the project is big enough to warrant one, researchers can have research "buddies" for smaller projects who advise and act as a bridge to the community, a patient panel can be set up to provide advice at different stages within a project, or workshops and focus groups can be run with patients and members of the public. In one U.K. project, we are currently using a combination of these approaches: we have a lay member on the project team, who also chairs the patient panel, updating the patient panel on the progress of the research and feeding back their comments to the project team. The patient panel has also nominated a member to sit on the project steering committee, so that these three project management groups—the project team, the patient panel, and the project steering committee—are all interlinked. These are all strategies that help incorporate research experiences and ideas from patients and the public, which can be highly novel and challenge the assumptions of researchers. Further information can be found online on the NIHR's INVOLVE website.[6]

1.4   SUMMARY

In this chapter, we addressed three important themes for the successful design and management of research in healthcare: ethics and governance, the benefits and pragmatics of research ethics review, and the involvement of people outside the research team in the design, conduct, and dissemination of research.

---

[6] http://www.invo.org.uk/resource-centre/resource-for-researchers/

CHAPTER 2

Readying the Researcher for Fieldwork in Healthcare

Heather Underwood, Ross Thomson, Anjum Chagpar, and
Dominic Furniss

This chapter is designed to help you prepare to undertake fieldwork in healthcare. Healthcare is an exciting and challenging domain where we might expose ourselves to events that are not experienced in normal workplaces or everyday life. These experiences could make you think about your own mortality and quality of life differently, you might have to face things you are not comfortable with, and this context might lead you to question or reaffirm what you think is valuable in work and life more generally. The work can involve sensitive and emotionally challenging situations. The work could also bring with it special considerations and rules to protect your safety and the safety of participants.

This chapter covers three broad areas focused on readying the researcher—researcher identity, emotion, and safety. These are intended to encourage self-reflection about important, but often under-acknowledged, aspects of performing fieldwork in healthcare. Considering these areas will help you prepare for a more successful research experience; it should also allow you to build a solid foundation that will benefit the research, your participants, and yourself. The content of the chapter is as follows:

How can reflecting on my role and identity help my work?	This encourages reflection on how you see yourself, your preferences, motivations, skills, knowledge, and limitations in a healthcare context. Reflexivity is introduced as a mechanism to think about how you impact the conduct and outcome of the research.
What emotional challenges will I face in fieldwork?	This encourages reflection on attachment and detachment toward your participants who could be in vulnerable and challenging circumstances; getting a grasp of what environments you would be comfortable within; and recognising how you deal with internal and external conflict.
How do I stay safe when doing fieldwork?	This raises safety concerns and procedures in healthcare that should be considered to protect yourself and your participants. This includes when you introduce a new technology into the healthcare setting.

2.1 HOW CAN REFLECTING ON MY ROLE AND IDENTITY HELP MY WORK?

When conducting field research in communities or in institutions, it is important to understand how you fit (and do not fit) within the research context, including the boundaries and biases associated with that role. This will help you understand how you interact with participants, how you collect and interpret data, and how you describe your research.

2.1.1 WHAT ARE YOUR PERSONAL AND RESEARCHER IDENTITIES?

Every so often it is worth taking some time to think about your personal and research identity so you have a clearer picture of who you are, what you are doing, and why. This should give you a view of the bigger picture and where you are headed. Are you doing healthcare fieldwork for a constrained one-off project, or are you making a strategic decision to develop a professional network, deeper expertise, and a career in this area? Or perhaps you are drawn to this area because you see it as one where you have the best chance of making an impact and helping others. Whatever your goal and circumstance, it is useful to understand it so you can choose to invest in activities that will payoff beyond the project, or rein in efforts where costs outweigh the shorter-term gains.

Healthcare is populated by people with different skills, expertise, and preferences. The differences can be stark: clinical settings such as hospitals versus non-clinical settings such as healthcare at home, healthcare for the body versus mental health, preventative medicine versus emergency aid, and so on. Even professionals with the same title can be very different, e.g., nurses who work in the operating theatre have very different skill sets, roles, and values compared to nurses who work in oncology. There is a very broad landscape in healthcare and many opportunities to engage with it. The choice of what and who to engage with might be shaped by external constraints, but perhaps also by internal goals, motivations, and preferences. You might be more comfortable and motivated to work in mental health, you might be drawn to a fast-paced, high-pressure environment like trauma surgery, or you might have a desire to help patients be more independent in their own homes. Examples of the variety of different projects and contexts, from an HCI perspective, are captured in the case studies in Volume 1.

Aligning your project with your personal preferences and motivations is more likely to lead to a successful outcome and a satisfying experience. Misaligning them could cause problems. For example, planning observations of surgical procedures while being squeamish might not be the best combination. There are other areas that the non-squeamish might find equally challenging for different reasons. There are different sets of questions you may want to ask yourself before starting project work in this area. If you already have a specific project in mind, ask yourself these questions: What are my short-term goals for this project? What are the long-term aims for my career? Do these align? Is there anything about the project that makes me feel uncomfortable? If you do not have a specific project in mind ask yourself these questions: What areas of healthcare really interest me? Are there particular patient groups that I would like my work to help, e.g., the elderly, children? Why does that interest me? Is there a research need in these areas? Who can I talk more about this with? Again, it is good to have a view of the bigger picture in terms of your research identity and career and how specific shorter-term projects align to this vision. Plans and identity can of course change and mature over time.

2.1.2   WHAT ARE YOUR ROLES AND RESPONSIBILITIES IN THE RESEARCH CONTEXT?

Some researchers will be entirely unfamiliar with the particular healthcare settings and conditions they are studying, whereas others will have knowledge and experiences that they bring with them into the health research environment. This knowledge and experience could be through working in healthcare or through more personal experience. Those who are unfamiliar with the context they are studying would benefit from visiting it to become more familiar with the place, procedures, patients, and staff. You will need to become familiar with: (1) getting to know the territory, and (2) learning what you need to answer your research questions (Lofland et al., 2006; Emerson et al., 1995). Before the study starts, it is useful to talk to practicing healthcare professionals, like nurses, occupational therapists, and different patient groups, to anticipate what you are getting into and the best approach to data gathering. This could be done before submitting or while waiting for ethical approval. Depending on the study and the resources involved, this could range from a few days to several months of voluntary work. This might sound like a long time, but this time will allow you to get a feel for the social and cultural practices and for others to feel comfortable with your presence there.

> **Tip:** Those who are unfamiliar with a patient group but who want to work with them can co-opt a patient representative onto their project so they can be an ongoing source of advice for the condition, lifestyle, and contacts. This is more commonly referred to as PPI (Patient and Public Involvement) in the U.K.—see Chapter 1.

In some situations time needs to be spent in learning practices and procedures so researchers, who are new to a particular context, are able to understand what they observe. For example, before studying paediatric cardiac and orthopaedic surgery Catchpole "spent 6 months preparing at the research hospitals by talking to practitioners, attending multi-disciplinary meetings, studying textbook descriptions of the operations, and by observing 38 pediatric and 10 orthopedic operations before data collection" (Catchpole et al., 2007). Organisational permissions will need to be sought to be present in context for learning purposes to prepare for research; however, no data gathering for research can be done prior to gaining the appropriate ethical approvals.

Tip: Depending on the topic of study, some researchers have spent months learning about medical procedures so that they are prepared for making sense of observations in a specialist context.

In all cases, there is normally a clear divide between your roles and responsibilities as a researcher, including where you can go and what you are allowed to do, and the roles and responsibilities of the healthcare professionals you are studying. For example, Furniss (Volume 1: Chapter 3) had a clear research protocol that meant he could not intervene with any medical procedure. However, this did not stop him helping out, e.g., by getting water for patients or fetching chairs for visitors; demonstrating willingness to help can assist with building rapport. Underwood (Volume 1: Chapter 12) took to comforting women who were in labour in her research context and talks explicitly about adding value by doing small jobs to help out in the research context. Similarly, Randell (Volume 1: Chapter 5) would play with the children when undertaking fieldwork in a paediatric surgical ward. Davis (2001) talks about more of these instances, and particularly how helping can alleviate feelings of marginality in ward life, build rapport with staff, and how partaking in ward life needs to be balanced with keeping one's distance. What specific individuals are comfortable doing, in different healthcare contexts, which also adheres to the conditions of their ethical approval, needs to be figured out on a case-by-case basis.

What you should do if participants ask for help or advice is something that should be considered in advance within the ethical protocol. Unless you are currently a registered professional (regardless of previous experience), or properly trained, it is best to avoid any ambiguities around identity and boundaries and not offer specific help or advice on medical matters. While this may be frustrating (especially if you feel you know the answers to participants' questions), you can signpost people to professionals, agencies, and organisations who are in a position to help. This is easy in clinical contexts, which have lots of professionals around. When working in the community, you can even carry relevant leaflets and contact information for local and national patient support groups, counselling organisations, self-help groups, independent benefit and social services advice services. The participant should always be advised to direct any specific health queries to their general practitioner or other clinician. On a related note, it may be beneficial (although not necessarily compulsory) to attend a basic life-support class so that in the event of a life-critical situation arising during a visit you are able to act swiftly until professional help arrives.

2.1.3   REFLEXIVITY: HOW DOES YOUR PERSPECTIVE IMPACT YOUR RESEARCH?

When conducting qualitative research you should be aware of factors that might affect your data, results, and conclusions. For those researchers taking a more positivistic stance, where they subscribe to the view that they are reporting an objective reality, it is more in-keeping to refer to "bias." A bias is some sort of deviation from the true picture. For those researchers taking a more interpretivist stance, where they believe that they are creating a picture from the data, "reflexivity" is a technique which should be used. For qualitative researchers, reflexivity facilitates a critical attitude toward locating the impact of research(er) context and subjectivity on project design, data collection, data analysis, and presentation of findings (Gough, 2003). Steier (1991) says, "… if researchers are to take seriously that knowledge is a social and cultural construction, they must apply this principle to themselves and their work. They co-produce rather than simply 'discover' the worlds of their research. Thus their own assumptions and activities as researchers must become part of the investigation within a process that explicitly tackles the complexities of multiple realities." In particular, this involves the relationships and rapport with participants, factors that impact the data collected, and methodological and theoretical perspectives that impact the interpretation of the data. Finlay (2003) notes that reflexivity has the potential to be a valuable tool that can help:

- examine the impact of the position, perspective, and presence of the researcher;

- promote rich insight through examining personal responses and interpersonal dynamics;

- open up unconscious motivations and implicit biases in the researcher's approach;

- empower others by opening up a more radical consciousness;

- evaluate the research process, method, and outcomes; and

- enable public scrutiny of the integrity of the research through offering a methodological log of research decisions (Finlay, 2002, p. 532).

Reflecting on how you affect and fit into the research context is important in healthcare because there are many different groups and individuals with different cultures, priorities, and values with explicit and implicit relationships to each other. These often hidden factors can reveal themselves negatively where we experience barriers and friction toward our work, or positively where things seem to happen almost too well—beyond expectations. Experiences of engaging and failing to engage with different groups are captured in the case studies of the companion volume (e.g., Volume 1: Chapter 7 shows great rapport with senior staff; Volume 1: Chapters 1 and 6 show difficulties with engaging with ward staff).

Figure 2.1: Ward Stories: Reflections on observing. The humorous film *Kitchen Stories* (2003) has inspired this illustration. The film encourages reflection on its tongue-in-cheek scientific rules for objective observation, e.g.: (1) build trust with your subjects, (2) collect data unobtrusively, (3) observe without interference, and (4) never become friends with your subjects.

Just a short while on a ward will highlight the roles of different groups and professionals in the workplace. Explicit uniforms differentiate some clinicians and their seniority. More implicit patterns of dress and behaviour indicate other groups, e.g., medics can dress smart casual, young medics might wear a stethoscope to show they are a doctor despite their youthful appearance, different colour uniforms distinguish nurses of different rank and healthcare assistants, and physiotherapists will generally look more sporty. People might also make assumptions about you, which can be influenced by how you look and what you are doing. While doing fieldwork in healthcare, Davis (2001) reports being asked to help with IT support by ward staff who knew she was studying technology, being asked for opening times from a frustrated patient who thought she was hospital staff, and being asked to help with medical procedures when wearing a sister's hat in theatre. Chapter 3 of this volume expands on engaging with different groups in healthcare more broadly.

When engaging with patients, there are factors that impact how you relate to them and how they relate to you, which can lead to potential biases in data gathering. For example, being identified or associated with the medical world can introduce a gratitude bias, especially in older people (Øvretveit, 1992). This is related to concerns raised by Rajkomar et al. (Volume 1: Chapter 11) who observe that interviewing patients can be a very delicate affair, as patients are reluctant to criticise the equipment they rely so heavily upon and do not want to appear less than competent with its use. Rajkomar et al. (Volume 1: Chapter 11) advise us to steer clear of more personal and subjective questions, e.g., "Is there anything you find difficult with the device?" and use more objective questions that stick to facts, e.g., "Could you tell me about any incidents that you have had with the device?" This is just one way of being aware of how our methods affect data gathering and subsequent analysis.

Another source of participant bias is the way that technology is introduced to participants. For example, in a study in India, Dell et al. (2012) found study participants were two and a half times more likely to prefer a technological artefact that they believed was developed by the interviewer, and five times more likely if the same interviewer had a translator. These effects are likely to apply to medical device evaluation too. In the same way that local partners could be co-opted to carry out interviews, patients could be co-opted to carry out studies (see PPI Section 1.3.1 in Chapter 1, this volume). One example, outside HCI and human factors, involves two people who suffered from bulimia interviewing each other to construct rich narratives of their experience (Tillmann-Healy and Kiesinger, 2001). Arguably, patients interviewing patients could reveal richer data because they have a shared experience that outsiders cannot empathise with to the same degree.

A recurrent theme in this section is the Hawthorne effect, which refers to how participants change their behaviour, perhaps unconsciously, in response to being researched. The Hawthorne effect gets its name from a study aimed to see whether changes in lighting conditions would impact the productivity of workers at the Hawthorne Works. They found that the act of being observed accounted for the changes in productivity rather than any change in environmental conditions per se. This effect is observed in other contexts. As explained above, this effect has been seen in participants not wanting to disappoint or criticise the work of researchers, but instead to please and meet the expectations of the researcher. The effect has also been seen in observational studies in healthcare, e.g., higher levels of compliance in hand hygiene procedures was observed when participants were visible to auditors (Srigley et al., 2014). This can be mitigated by keeping participants blind to the true focus of the study, by being present in the context for a long period so participants become used to your presence and behave more normally, and clearly explaining that the participants are not being tested but the system or device is being tested in the case of doing field tests.

Similarly, biases could be introduced not only in the data gathered from participants but also in interpretation of the data. For example, Underwood (Volume 1: Chapter 12) reminds us that we need to be open to criticism about our own design rather than being dismissive of others' views. Again, using an independent third party to help with evaluation could reduce the problem of evaluator bias.

Reflexivity could also consider how the gender, race, condition or other distinguishing factors of both the researcher and the participants affect the research process. Would a man get the same results if interviewing a woman about an app to support her menstruation cycles? Would someone with diabetes get more fruitful data if interviewing other people about devices for diabetes management? Reflexivity should include any significant effects from the researcher's background and theoretical perspectives that may have influenced data gathering and analysis. Such effects are often unavoidable, but it is important to be aware of their likely impact on the research findings as far as possible. Detailing these effects should not be seen as listing limitations and weakness with the work, but rather providing a more complete picture of the potential influences on the research process and outcomes.

For more on reflexivity: Finlay (2002) argues that reflexivity is a valuable tool for building trust and integrity in qualitative research. She describes different stances toward reflexivity and how it can be used at different stages of research. Barry et al. (1999) describe how reflexivity can be used to develop research and enhance collaboration in a multidisciplinary team. Finlay and Gough (2003) have edited a volume that shows many different perspectives, stances, and examples of researchers using reflexivity.

2.2   WHAT EMOTIONAL CHALLENGES WILL I FACE IN FIELDWORK?

The emotional nature of qualitative research includes how it affects people who are involved in the research and "emotions" as a subject of study in its own right (e.g., see Gilbert, 2001). This is particularly important in healthcare where situations are emotionally charged. For example, Volume 1 contains a number of incidents where researchers were disturbed by their experiences (e.g., see Volume 1: Chapters 2–5). Gilbert (2001b) describes how sometimes researchers can be "invisibly affected" by events (e.g., shown through flashbacks, dreams, and withdrawal from research activities) and argues for more effort to recognise and deal with these situations. Physically leaving the field setting is easy; emotionally getting away from the field can be more challenging (Wincup, 2001).

2.2.1   WHAT IS YOUR PERSONAL LEVEL OF EMOTIONAL ATTACHEMENT OR DETACHMENT?

Different studies will require different levels of engagement with staff and patients. It can be difficult to be invited into people's lives, participants' homes, or a healthcare institution, and not become emotionally attached to your participants. While some research methodologies require the researcher to maintain the role of an objective observer, across the spectrum there are other methods that acknowledge the importance of engagement and encourage the researcher to enter into the participant's "lifeworld," developing empathy, trust, and rapport (Smith et al., 2009). At the far end of this spectrum, the interpersonal relationships that can be formed utilising these latter methods in healthcare, when people are potentially at their most vulnerable, can mirror that of the counselling or psychotherapeutic relationships. This can be emotionally demanding at best and requires great consideration early on in the research process because it is so involved and requires very particular skills. Generally HCI and human factors researchers are not trained counsellors and so such methods should only be used when collaborating with people with the appropriate experience and skill set. There are huge ethical issues here about managing participants' expectations, not going beyond our expertise, and withdrawing from the research context (which is dealt with below). Even toward the middle of the spectrum there is debate about how participant relationships are managed so rapport is not just a means of gaining richer data but so there is "a less exploitative relationship based on informality, equality, reciprocity, empathy, rapport and subjectivity" (Wincup, 2001). Thought should be given to the nature of the researcher-participant relationship, and a balance needs to be struck between getting too attached and being too detached (Gilbert, 2001a).

Strategies to help decrease emotional stress or "burnout" have been well documented in the nursing and counselling literature and may be relevant to researchers spending in-depth time in the field. For example, Watkins (1983) describes the importance of different ways of preventing or overcoming emotional burnout such as making time to engage in physical (as opposed to cognitive) activities or scheduling periods in which one can completely disconnect from the emotional world of others. Some people may use self-reflection and journal keeping as a way of working through some of the emotions that are brought to light (Gilbert, 2001a), while others may use the counselling services provided by most universities and healthcare institutions. One example of how to write a reflexive journal is given by Saunders (2002).

> **Tip:** Establish strategies for dealing with emotional stresses during your research—journaling, weekly phone calls to a friend or family member, exercise, etc.

The process of ending the participant-researcher relationship after the study can also be an emotional time that is worth considering. Some of these issues have been eloquently described by Cutcliffe and Ramcharan (2002) and Booth (1998). For example Cutcliffe and Ramcharan offered participants the chance to review reports and attend seminars intended to disseminate results. This approach allowed participants to feel that they had not experienced a "hit-and-run" interview, and Booth (1998) describes how she believes that conducting participatory research with people with learning difficulties is not only about inclusion but also about a longer-term commitment. Her work with a particularly lonely and sensitive group of women has left a legacy of being in contact with some of the individuals 11 years after the study. Withdrawing from a study after giving participants a prototype or intervention is addressed in Chapter 5, this volume.

Emotions can also run high in studies where there is little direct interaction with participants, but you are affected by what you hear and observe. Gilbert (2001b) describes researchers being disturbed by a transcription task as she had not appreciated how they would be affected by the content of the interviews. These emotional effects can also change over time as you get emotionally attached, detached, or challenged by the context that you are engaging with. For example, Furniss (Volume 1: Chapter 3) was worried about starting a study on an oncology ward and how to respond and react to the patients there. Over time he became more comfortable as he accepted that he could do little to help the patients personally, that his role was limited to research, and that the patients were being cared for as well as possible by the healthcare professionals. He became less emotionally involved. In contrast, Chagpar (Volume 1: Chapter 4) was carrying out a study in surgery and became increasingly disturbed by not seeing the patients after they had been put to sleep and had their surgery. She realised that she was attached to these patients and wanted to know their outcome. Staff members build coping strategies as part of their work, often over many years. As HCI researchers there is an acute need to recognise what these strategies are for us, and to develop them quickly for research projects that are time limited.

2.2.2   ARE YOU COMFORTABLE WITH BLOOD, DEATH, BIRTH, ILLNESS, NUDITY, SMALL SPACES, PROVIDING EMOTIONAL SUPPORT...?

Observing and working with people who are in vulnerable situations on account of their health can be uncomfortable and unsettling. These feelings are particularly intensified when conducting research in participants' homes, where unfamiliar things surround you. Some places may be so clean that you are afraid you may mess something up; other places may not have been cleaned for years. You may visit people who are unable to get out of bed. You may encounter alarming smells, sights, and sounds. In order to be sensitive and compassionate in these situations, it is advisable to try to anticipate what you are comfortable with in *advance* of conducting your research. While doing research in community settings is varied and unpredictable, there are things that you can do to make it more comfortable for yourself. For example, if the sight of blood makes you ill, do not visit someone on home dialysis during treatment. Being unprepared for the situations in which you will interact with your participants can lead to embarrassment for you and your participant and, furthermore, negatively impact the quality and quantity of data you are able to collect.

It is important to be aware of, and reflect on, one's strengths and vulnerabilities relative to particular health situations. There are many different contexts and so sometimes adaptations to research plans can be made; e.g., one of the authors was not comfortable with observing surgery and so research efforts (understanding the use of infusion devices) were directed more toward ward areas. Different environments can be challenging in different ways; e.g., ICU can be peaceful and involve little interaction with very critically ill patients because they are often unconscious; ward areas can involve less acute dangers for patients but there might be more emotion as people are communicative; and areas such as mental health may involve working with people who are suffering from depression, schizophrenia, or another condition that makes engagement challenging. There is a rich diversity of healthcare settings where important contributions can be made.

The researcher's perceptions can change with experience. In the very earliest days it is advisable to speak to someone experienced in the healthcare setting you plan to study and have a few days there to check that you are comfortable with it. After days, weeks, and months you may find that your perceptions change. For example, worries about being exposed to people in vulnerable situations may subside, your passion to help people may be energised and reinforced, or contrastingly you may develop areas of discomfort that were only marginal or that you did not predict before the start of the study. Be aware of your own journey and have a support network to help you along it. Frequent opportunities to debrief with an experienced professional either in research or in the domain (e.g., a nurse or hospital employee), particularly during the early phases of a study, can help you reflect on your experiences.

2.2.3   HOW DO YOU HANDLE CONFLICT?

Carrying out research in the community and in participants' homes (as opposed to in a healthcare institution) will hopefully encourage participants to speak and act in a more natural way, as they will probably feel safer in a more familiar environment. While this may be advantageous for data collection, it does mean that you may witness behaviours or attitudes that could result in some internal conflict about your research participant. While you want the person to engage in an in-depth interview about a healthcare issue, you may find that the participant holds views counter to your own or behaves in a way you disagree with. For example, imagine visiting a patient with lung cancer who still smokes, or makes a racist remark. Situations in healthcare institutions may also give rise to conflict. For example, imagine someone who is in surgery has committed a terrible crime and is prioritised over other needy patients. You might have mixed or conflicted feelings about these and similar sorts of situations.

Clearly, making your disapproval known will result in a conflict that could threaten the integrity of your data and your relationship with your participants. Even your nonverbal behaviour may reveal your inner feelings toward the person or situation. One possible solution is to adopt a technique used in counselling whereby you separate the person from their behaviour. This is known as unconditional positive regard, and it is a way of viewing the person as valuable without either condemning or condoning the person's actions and views (Rogers, 1962).

According to Marshall (2006) people can respond to conflict in different ways depending on the circumstances, e.g., by accommodating, avoiding, collaborating, competing, and compromising. She promotes education and training in communication skills and conflict management as a way for nurses to become better equipped to manage problems as they arise. Learning to overcome obstacles and resolve conflicts will also allow you to successfully conduct your study, and will ultimately benefit both you and your participants. Objectifying conflicting or upsetting experiences can be made easier in the process of writing fieldnotes and is another reason why they should be written soon after each session (Bernard, 2002; Berg 2004).

2.3   HOW DO I STAY SAFE WHEN DOING FIELDWORK?

You should consider the safety of yourself and your participants before conducting any fieldwork in healthcare.

2.3.1   WHAT ARE THE RISKS TO YOUR SAFETY?

One advantage of working in hospitals is that you will be around many healthcare professionals who can guide you in what is and is not safe. Sometimes the safest thing to do is to shadow a healthcare professional who can act as a chaperone if they have time. However, this will not be possible in all contexts. Initial meetings with professionals who work in the situation, or who at least are familiar with the medical context, can relieve anxiety for you and for them, e.g., they will be concerned with your safety and how you might impact the safety of others on the ward. It is advisable to at least have a safety briefing from someone experienced in that context to outline any safety issues you should be aware of for yourself, staff, and patients. For example, working on a haematology ward, an area where infection control is important, it was necessary to be familiar with the rules and procedures for patients who were infectious or vulnerable to infection. Signs on doors in the oncology ward also indicated rooms that were radioactive as patients who were receiving radiotherapy were there or had recently stayed there. We were also advised to wear short-sleeved shirts and not to have any jewellery below our elbows for infection control purposes. Regularly washing hands in sinks and with alcohol gel is expected of all staff and visitors in most modern hospitals to help reduce the spread of infection.

Thomson et al. (Volume 1: Chapter 10) emphasise that conducting research in the community and participants' homes puts the responsibility for safety firmly in the hands of the researcher themselves. Travelling to unfamiliar areas by car or on public transport requires much planning and ideally should take place during daylight hours. Similarly, when out and about it is advisable to keep laptops and mobile phones out of sight, as these can be a target for street crime in some areas. There may also be risks to your safety inside a participant's home. These may range from smoking to troublesome pets. If you feel uncomfortable around a participant's dog, ask if they can be kept out of the room for the duration of the visit. Ask a participant who smokes if they can open a window. These issues may have to be handled with diplomacy, but remember you are responsible for your own safety. It might be advisable to travel with a colleague—this can also be useful for interviews where one person speaks and the other takes notes. If your organisation does not have a lone-worker procedure, it is worth setting up a system whereby someone knows where you are going, what time you expect to be there and for how long, and that you call your supervisor or other designated person when you have finished your visit to confirm that you are safe.

> **Tip:** Trust your instincts—if you feel in any way threatened or uncomfortable, it is perfectly reasonable to stop what you are doing and exit the situation. Things can always be rearranged if needed.

The safety concerns related to working in health centres in low-income regions or developing countries are intricately linked with the specific cultural, social, and environmental hazards present in these settings. Standards of care, including sanitation protocols, can vary dramatically from facility to facility, which can lead to an increased risk of exposure to contaminants and/or country-specific communicable diseases. In addition, it is important to have a basic understanding of the cultural norms in the facility you are working in and expectations regarding dress and behaviour. In strict patriarchal societies, women should be especially cautious and aware of the cultural and societal norms expected of women. Even if you do not agree with or condone these norms, abiding by them while conducting your research may be essential to ensuring your safety and the safety of your participants.

2.3.2   WHAT ARE THE RISKS TO THE PARTICIPANTS' SAFETY?

You have responsibility to ensure that your research does not cause any negative consequences for participants. The ethical approval process checks risks to patients inside and outside the hospital. In the community, if there are no healthcare professionals present, there can be more onus on the researcher to make the participant feel safe while taking part in your study. Responsibility to provide safety beyond the research project needs to be kept in perspective, and should be considered by the research team. Things can be simpler in hospitals as there are normally trained and experienced professionals within easy reach.

In the U.K., NHS ethics committees will usually require you to undergo a Disclosure and Barring Service (DBS, formerly Criminal Records Bureau) check, which helps assess whether you may be a risk to certain vulnerable groups. This is often needed for working with patients inside and outside of hospitals. For outside hospital, Thomson et al. (Volume 1: Chapter 10) suggest that even if your ethics committee does not require this, the ability to produce such a document may help vulnerable participants feel more secure in interacting with you. On a similar theme, if you are intending to meet with participants on a one-to-one basis, it may ease any anxieties they have by offering them the option of having a friend or family member present in the house or arranging to meet them in a public place to conduct an interview.

With regards to participants' actual safety, it is worth remembering that older people in general as well as people with certain illnesses are less capable of fighting infections. With this in mind, it may be worth carrying some alcohol gel hand wash with you on visits to people's homes (this may also be used to protect yourself) and bearing in mind that you may have to cancel meetings with participants even if you are feeling only slightly unwell to avoid passing on any illness. For example, an older person with Chronic Obstructive Pulmonary Disease would not want to interact with someone who may have a slight cough or cold. Similarly, visiting wards if you have diarrhoea or even just a cold should be avoided.

In many cases research protocols will dictate that you cannot get involved in delivering medical care. However, one of us has worked in areas where existing medical care is scarce and where the number of highly skilled medical professionals is low (Volume 1: Chapter 12). In these situations it can be tempting to provide basic medical care to those in need, especially if you have prepared yourself by taking basic life-support classes as mentioned earlier. However, it is extremely important to know when you may be overstepping your medical qualifications, or discrediting the skills of the healthcare providers by intervening in a medical case. Performing medical care under any circumstances can place you and the patient at risk, and it is important when entering a healthcare environment as a researcher to make it clear what qualifications and expertise you have (and do not have), and to set clear expectations about the work you are there to do.

2.3.3   WHAT ARE THE HEALTH RISKS INVOLVED WITH A TECHNOLOGY INTERVENTION?

Interdisciplinary collaboration is very important when introducing a new technology into a healthcare setting (this is covered in detail in Chapter 5, this volume). Often, technology is not designed by or with healthcare professionals, which can lead to an inappropriate or unsafe result. For example, the case study that looks at digital pen technology in labour wards (Volume

1: Chapter 12) did not initially consider the possibility of the pens being a hazard for spreading infection. The small seams and cracks in the pen casing could carry germs from one patient to another if not disinfected between each patient interaction. The nurses in this study mostly followed a hand-washing policy between patients. However, when the nurses began using the pens, a protocol for disinfecting the pens between each patient was not established immediately, and only came up later during a discussion with hospital administration. Fortunately, there were no negative health consequences attributed to the technology intervention in this case, but this illustrates the importance of collaborating with healthcare staff about the exact specifications and design of a new technology *before* implementation.

If you are involved in introducing technology into the home environment, be aware that, unlike the controlled hospital environment of the hospital or clinic, the home is much more variable. Some homes can be particularly damp or dusty or have an unreliable/variable power supply and so electrical technologies may become a fire hazard. Sockets in homes are not always in the most convenient places and so trailing leads may be a trip hazard. Researchers should be aware of these and other potential hazards in the study environment.

2.4   SUMMARY

In this chapter we introduced three important themes for readying yourself for fieldwork in healthcare: identity, emotion, and safety. This advice could apply to fieldwork in other problematic contexts, which have the potential to be unusual, emotionally uncomfortable, and where safety could be a concern, e.g., studying inmates in prisons or soldiers in a conflict zone. Healthcare is in some sense remarkable because of the work it does in preserving and enhancing the quality of life of the most vulnerable. However, it is also unremarkable in the sense that we rely on these services on a frequent basis. The researcher should not assume they are entering a normal place of work and instead prepare themselves for their area of study. The theme of researcher identity is designed to encourage you to think about your own role within the research context, any biases that may affect your research, and the importance of identifying and recognising boundaries related to your work. Within healthcare, and even within a single hospital, there are very different types of research context that have different treatment, care, and challenges to face. The section on emotion encouraged you to reflect on possible conflicts you may come across in your research as well as issues related to the relationships you form and emotive situations you encounter during your research. Healthcare caters to people in very vulnerable situations from all walks of life. Witnessing these situations and the wide variety of people and behaviours can challenge you, and bring joy and despair. It could even lead you to think about your own life and mortality differently. The subject of the final section is that of safety. It is designed to encourage you to consider not only the safety of participants, and risks introduced by technologies, but also your own safety. Conducting fieldwork requires you to constantly assess risks (to both yourself and others in the context).

The importance of planning and design for the success of a research project cannot be understated; preparing yourself is also part of that process. While not intended to be prescriptive or exhaustive, we hope that working though this chapter will have encouraged a better understanding of some of the more personal issues involved in healthcare fieldwork. Taking the time to "ready the researcher" will help you, your colleagues, and your participants move toward a more successful and satisfying research experience.

CHAPTER   3

Establishing and Maintaining Relationships in Healthcare Fields

Svetlena Taneva, Aisling Ann O'Kane, Raj Ratwani, Brian Hilligoss, Anja Thieme, and Kristina Groth

The success of any fieldwork research depends on the help of a number of people. This includes those who provide opportunities for access to a research site and for recruitment of specific participant populations, those who participate in the study, and of course the research team. HCI and human factors research in healthcare presents intricacies and complexities that can be hard to navigate for researchers without experience in this domain. As seen in the case studies contained in the companion volume, developing working relationships that are built on trust and mutual benefit can be difficult, but they can also make or break a research agenda in the healthcare space. This chapter aims to prepare the researcher for what they should consider and may encounter when establishing and maintaining relationships in healthcare. This includes the review of healthcare dynamics, negotiating access to the field, establishing relationships with healthcare staff, and maintaining relationships with patients and caregivers. The goal is to prepare you for the social aspects of healthcare fieldwork that may influence your research.

The table below names and describes the content of each section.

What healthcare dynamics may affect my study?	Many healthcare settings are characterised by complex organisational structures that may influence building and maintaining relationships during your research. Hierarchies and subtle differences in healthcare settings are reviewed.
How do I negotiate access to the field?	In order to get access to healthcare settings, an entry point needs to be identified and negotiated, either through organisational means or through individuals such as key enablers.
How do I establish relationships with healthcare staff?	To ensure your project will be well received by clinical staff, it is important to establish your presence and identity in the clinical setting. Finding a "champion" and how you present your project and your recruitment strategy for clinician participants are also critical factors.

How do I manage relationships with patients and caregivers?	Patients and caregivers can provide rich information about healthcare contexts for your research, but relationships can be hard to attain and maintain. Issues with recruitment, data collection, and empathy with patients in clinical and non-clinical settings are highlighted.

3.1   WHAT HEALTHCARE DYNAMICS MAY AFFECT MY STUDY?

The organisation and delivery of healthcare, and by extension the cultures within healthcare settings, can vary considerably from site to site, and from one country to another. There are, however, some fairly common structures and dynamics across settings that come with implications for establishing and maintaining effective relationships with a field site. Among these are the varied mix of professions, subdivisions within each profession, hierarchies, time sensitivities, types of clinical settings, and the broader context of healthcare. This chapter focuses on the social context and dynamics of healthcare settings.

3.1.1   WHAT DO I NEED TO KNOW ABOUT THE DYNAMICS WITHIN THE VARIED MIX OF PROFESSIONS?

The delivery of healthcare involves a variety of professions: doctors, nurses, various technicians, therapists, other allied health professionals (e.g., respiratory therapists, emergency medical technicians, imaging technicians, etc.), clerical staff, and administrators are all essential parties. Each profession has its own skill set, and many require specific training and licensure. Clinicians and other professionals identify very strongly with their profession and due to their different values and roles important inter-professional dynamics play out in practice. For example, while doctors tend to wield considerable power in healthcare settings through clinical expertise, administrators often exert power through budgetary control. Consequently, tensions between the clinical and administrative realms can exist. In the U.S., Canada, and other places where doctors are not employed by the hospitals or health systems, administrators have limited power to shape or alter practices—rather, it is doctors who have a great degree of control. This is due to their independent standing with respect to a hospital—if a well-respected doctor is unhappy with specific hospital practice policy, he or she may decide to switch hospitals, leaving the original institution without a valued doctor. Awareness of such dynamics can help a researcher better interpret situated observations of patient care.

3.1.2   HOW DO CLINICAL SPECIALITIES IMPACT MY RESEARCH?

Within each clinical profession there are various subdivisions, such as internal medicine, gastro enterology, and surgery. Each entails a different approach and skill set for providing treatment.

These specialties, in turn, are further subdivided into sub-specialties (e.g., cardiology, nephrology, thoracic surgery, etc.). With these various sub-specialties come differences in terminologies, technologies, sub-cultures, and orientations toward the diagnosis and treatment of illnesses. Tensions between these various specialties and sub-specialties emerge from these differences and from the tendency to identify closely with one's own sub-specialty. Turf wars and concerns over territory are not uncommon—for example, emergency department doctors and doctors on inpatient services may have opposing views on the appropriateness of a patient's admission to hospital (e.g., Volume 1: Chapter 6). A new researcher would need to examine such perspectives from all angles in order to get a holistic understanding of the situation, while maintaining sufficient neutrality so that individuals would be comfortable opening up and sharing their thoughts freely.

3.1.3   WHAT IS THE ROLE OF HIERARCHIES IN HEALTHCARE?

Healthcare is characterised by steep hierarchies. Doctors tend to rank above nurses, who in turn rank higher than technicians and clerical staff. Within each profession there are often hierarchies as well. For instance, in Internal Medicine in a Canadian hospital, an attending physician supervises fellows and residents, while senior residents supervise interns and other junior counterparts. Similar hierarchies exist in other countries, although the names for the positions are different (e.g., consultants, specialist trainees, and foundation year staff in the U.K.). These hierarchies highlight differences in terms of experience, seniority, and power. One implication of hierarchies is that sometimes those at the lower levels tend to be assigned the least desirable shift schedules. There is also evidence that these hierarchies affect patient safety, since parties at lower levels may be reluctant to speak up, contradict, or challenge their superiors (Walton, 2006).

3.1.4   WHAT DO I NEED TO KNOW ABOUT THE BROADER CONTEXT OF HEALTHCARE?

All of these dynamics happen within a broader context of continued flux. Healthcare in many countries today is undergoing considerable change. Currently, this is notable in the U.S. where the first major reform of healthcare in more than forty years is happening by way of the Affordable Care Act. There, efforts are underway to take a system that is heavily focused on specialty care in acute settings and transform it into one that is patient-centred and focused on primary and preventative care (Koh and Sebelis, 2010). Around the world, pressures to reduce costs while improving the quality of and access to health care services are mounting and healthcare organisations are experimenting with a variety of approaches. In Sweden, for example, four layers of healthcare will be implemented in the near future: primary care, specialist care, emergency care, and highly specialised care. Such approaches put high demands on collaboration between different care levels. While telemedicine has been available for a long time in Sweden, specialised care follow-up through telemedicine programs is starting to be rolled out across remote areas in Canada. For you, the fieldworker, the message is to expect a changing, perhaps even volatile context.

3.1.5   WHAT ARE THE DIFFERENT TYPES OF CLINICAL SETTINGS?

There are differences in the settings where care is delivered. Inpatient or hospital settings are focused on addressing acute illness episodes. Outpatient or ambulatory settings are focused on addressing chronic and less severe episodes of illness. Both inpatient and outpatient settings may also be differentiated in terms of whether or not they entail a teaching mission. Academically affiliated hospitals, for example, provide opportunities for clinical experience and training for both students and recent graduates of medical, nursing, and other health sciences training programs. With missions that include education and sometimes research, in addition to patient care, these settings are often accustomed to and the most accommodating of research endeavours. On the other hand, smaller hospitals that do not have many research projects might be quite appreciative of the attention and opportunity to contribute to research, and consequently conducive to fieldwork research agendas. You should take these factors into account when selecting study settings, and recognising these factors may also help you to make sense of things that occur during data gathering.

3.1.6   WHAT IS A HOSPITAL SITE LIKE?

Because healthcare, at least in the hospital setting, is a 24-hour, seven-days-a-week endeavour, there is an important temporal dimension to consider. Much of healthcare in such settings is accomplished by shift work and rotations, the durations of which can vary considerably from one unit to the next or from one profession to another. In large, highly specialised hospitals, the sheer volume of staff, combined with continual rotations, can result in an ever-changing mix of personnel. This might mean that some staff have little or no prior experience of working together. This, combined with the need to better integrate care for complex patients with multiple conditions, has generated a shift from an emphasis on care delivered by individual clinicians to care delivered by inter-professional and inter-specialty teams. The environment in large hospitals is in constant flux, where staff often change, policies evolve, and roles and responsibilities adapt to the requirements of the situation. For example, one will often see a nurse show a new resident physician on his first day in the hospital how to use the electronic health record system—something that is generally considered out of scope of the nurse's responsibilities. Such factors of continuous change can make it difficult to develop rapport with staff, and sometimes make it difficult even to work out people's roles and status.

Patients and family members are likely to be present to different degrees at different times of day depending on the clinical context. This requires proper conduct on the researchers' part. It also might change the dynamic of things like gaining consent, interviews, and observations.

Tip: Be aware of the 24/7 nature of some clinical settings and decide when would be a good time to visit the site. There might be different activities, people, and pressures at different times during the day.

3.2 HOW DO I NEGOTIATE ACCESS TO THE FIELD?

3.2.1 HOW DO I GET MY FOOT IN THE DOOR?

Many hospitals have a group that provides formal access for research and development within the hospital, to allow researchers to get their "foot in the door." Such entities could be a department or office within the hospital that has been given the responsibility to manage all contacts with industry and academia, and to track research projects within the hospital; or it could be a research institute closely connected with the hospital. In U.K. hospitals, this is generally the Research and Development (R&D) Department. These formal entry points should be used in order to identify key study enablers who can provide access to the field.

Some hospitals or healthcare organisations may not have the same tradition in research and so they may not have a formal department to facilitate access for research purposes. To identify key study enablers for such settings, you might consider contacting other researchers who have already gained access to the intended hospital or clinical setting (e.g., Volume 1: Chapter 6). Alternatively, you could identify and contact clinicians or other staff in the hospital whose background indicates that they may be interested in your research area.

3.2.2 WHO COULD BE A KEY STUDY ENABLER FOR MY RESEARCH AGENDA?

There are two main strategies for establishing links with a specific healthcare environment: top-down or from the frontlines. Top-down focuses on the managers, starting from the top level and moving down the hierarchy. Potential targets for contact could be either clinical managers or IT managers, such as the hospital's Chief Technology Officer. The managers themselves are usually not the key study enablers; instead they are the first step to finding that person. The path may not always be straightforward, but the advantage with the top-down approach is that management approval is secured first, facilitating access and potentially buy-in downstream. Once an enabler at the management level is secured, you need to plan out, with this individual, your entry strategy to the frontline staff.

Figure 3.1: Dissimilar groups: Access and study. Accessing and studying a group very dissimilar to ourselves can be difficult. Strategies to help include: access via a gatekeeper, provide help if possible, research something of value for that group, and offer some incentive to participate.

Starting from the frontlines means establishing contact with someone who works as a care provider in the specific clinical setting. Ideally, this would be an individual who has shown an interest in the specific research. For example, a key study enabler at the frontlines can be a nurse with responsibilities that can be helpful when conducting the study. This person could be identified through networking activities or online research. In one author's experience, an expert surgeon was identified as a potential study enabler through his outspoken interest in achieving a change in how he was working with patients and collaborating with other surgeons remotely. After the contact with the key study enabler is established, then approval from the clinical management is usually needed.

It is important to explain the contribution of qualitative research to your contacts; research in healthcare has a tradition of being strongly quantitative and some healthcare professionals may be unfamiliar with or even quite negative about the use of qualitative methods. One good strategy to overcome such resistance is to discuss a recent qualitative study that applies similar methodology and was published in a high impact journal (e.g., the *Journal of the American Medical Association*). Alternatively, you could refer to the work of your colleagues in other clinical settings and the value it brought to improving quality, efficiency, and safety. Sometimes it can be difficult to change people's opinion on qualitative research and you may have to find another key study enabler. The key to success is that the enabler should believe in what you are doing and what you will achieve.

> **Tip:** Network with researchers, managers, and frontline staff who have connections to potential field sites and be open to opportunities that arise. Be aware that people have different interests and biases when it comes to research and try to use this knowledge to facilitate collaboration with mutual benefits.

3.3   HOW DO I ESTABLISH RELATIONSHIPS WITH HEALTHCARE STAFF?

3.3.1   HOW DO I MAKE MYSELF KNOWN IN THE INTENDED STUDY ENVIRONMENT?

Successfully conducting fieldwork in healthcare requires understanding the dynamics between clinical staff and determining how you, and your research, can fit into the environment. As soon as you know that you want to conduct a study in a particular clinical environment and have negotiated access to the intended site, you should try to be present in that environment as much as possible. Make frequent visits to the clinical environment and stay in the environment, get to know the staff, and informally observe. Your presence will allow you to build trust with the staff and the staff will become accustomed to your presence, so that when the time comes to conduct your study, staff are comfortable with you. If you are able to engage particular staff members in aspects of your study planning, such as the study design, there will be greater staff engagement when you conduct your study. Use the opportunity of being there to get an understanding of issues that staff care about; find a way to weave these into your research plan, even if as a side project—this will make your study all the more relevant and staff will see benefit in participating in your study. The more the staff feel part of the study, the more committed they will be to seeing your study succeed—make time to explain how the study will be relevant to them. Your study may address an issue that staff are currently trying to solve and the topic might be of deep interest to them. In one case, the researcher discovered this quite late during her official project presentation to staff—one clinician approached her, saying it would have been beneficial if she had given the presentation when she first started visiting the department—six months earlier (Volume 1: Chapter 2). The staff may have suggestions for data collection or access to certain databases that may contain information that would facilitate your study. In addition, they can provide valuable input for ethics applications (e.g., regarding appropriate consent practices). In one instance, a study site withdrew from an ethics board-approved study because staff felt the chosen consent process did not fit well with their environment and process (Volume 1: Chapter 5).

In addition to building trust and having staff grow accustomed to your presence, it is important to observe the dynamics between the staff in the environment. You may notice that some staff do not get along with other staff members and it will be important to understand these dynamics to achieve the goals of your study. Spending time understanding the culture of the clinical environment will pay off in the long run. Once you have spent some time in the environment and feel comfortable with the staff and understand their interests, you should begin to identify what role the staff can play in your study. Staff who are highly engaged might serve as collaborators and even co-investigators. Other staff or patients may be participants in your study. It is important to make sure you have at least one clinical "champion" (see Volume 1: Chapter 3). This is somebody in the clinical environment who is interested in seeing your research project succeed. Without such a champion, it can be very difficult to conduct a successful study. Note that the project champion is a different role from the project enabler, although they may be the same person. The project enabler facilitates access and connections, particularly at the beginning of your study. The project champion is involved throughout your project and facilitates the day-to-day work of your study.

3.3.2   HOW DO I PRESENT MY STUDY?

It is important to present the study protocol to potential participants as early as possible. This will often be after ethics approval (see Chapter 1) has been obtained, but it may be useful to do so even earlier so that study participants have a real opportunity to shape the proposed study. The format, content, and entry strategy for this initial presentation are critical factors to gauge interest and encourage participation in your study. Your target participants may include one or more professional groups (e.g., doctors, nurses, allied health professionals, and clerical staff). Given their busy schedules, you have to identify an efficient way to present the study details to these potential participants. It is not uncommon that the clinics at university hospitals have a program for presenting research projects, for example at morning meetings, in-service meetings, or seminar talks (e.g., a "lunch and learn").

You should also be prepared for informal one-on-one presentations where you can give a 30-second "elevator pitch" of the study in the hospital hallways, as well as for more formal presentations at internal meetings. Part of constructing your identity and defining your role in a project is being able to effectively communicate the goals of the study and the value of the work. In presenting the study goals and methods, you should refer to issues in the environment that you identified through your initial informal observations and explain how your study will address those. In every project there are multiple stakeholders, and the way you present your ideas and contributions will be largely dependent on your audience. On the other hand, you will certainly need to engage staff in a bit of education about HCI and human factors philosophy—mainly, you should emphasize that the focus is on system-level factors and removal of blame from individuals (e.g., see Volume 1: Chapter 3).

Using the appropriate vocabulary is a key component to effectively communicating what you do to a variety of audiences. While explaining your work to other researchers and healthcare professionals, it may be appropriate to use technical terms and phrases that are part of their routine vocabulary (e.g., medical terms). However, when working in the community and with patients, using formal or "jargony" language may put people off, whether it is medical or HCI research language. Also, the same words can mean quite different things in different contexts, which can lead to communication mismatches that may not be discovered until much later in the project. For example, in the U.K., we referred to "incidents" like we would "events," i.e., meant in a broad sense like its use in the critical incident technique (Flanagan, 1954). However, healthcare professionals said "incident" was synonmous with a serious medical error in healthcare, which was not our intention or the focus of our work. The use of language should be tailored to the participants' capabilities and life experience in a bid to reduce power differences that may otherwise encourage a lack of authenticity and disclosure between researchers and participants (Karnieli-Miller et al., 2009).

The presentations should be delivered on several occasions, to ensure you reach staff who work various shifts. You should also consider any incentives to increase the likelihood of attendance to your presentations, such as providing breakfast or lunch during the presentation. Most importantly, you should discuss your entry strategy for the study presentation with your study champion within the department and with the clinical managers—these individuals will not only provide valuable insights on what approaches will appeal to their staff, but will also facilitate the distribution of the invitation to your presentation.

> **Tip:** Try to present your research questions and study design as early as possible at internal clinician meetings such as in-services, continuing education sessions, or seminars. Compose several versions of your "elevator pitch" to describe your intended work "on the go"—each version should target a particular audience, using appropriate vocabulary.

3.3.3   HOW DO I RECRUIT CLINICIAN PARTICIPANTS?

Participant recruitment is a crucial aspect of the success of fieldwork studies in healthcare. Participant recruitment could be facilitated by a local site coordinator, but the coordinator would likely need to be compensated for their time. He or she could send out the initial information packages, informally speak to staff about the study, arrange your recruitment presentation, send out reminders for deadlines, forward questions and concerns to you, etc. If you don't have these formal arrangements, a study champion could advise you or help you voluntarily in a less resource-intensive way. The success of recruitment could also be influenced by how you choose to obtain consent from participants. You need to consider a number of challenges associated with each strategy. Where there is a lot of involvement from participants in recruitment and other research work, for example in participatory design projects, providing funds for healthcare participants to "buy" time from their clinical work can be a great advantage (Volume 1: Chapter 7).

Unique factors to the healthcare context that can affect recruitment include culture, busy schedules, fear of litigation, and avoiding potential accountability. To mitigate the effects of these factors, researchers conducting low-risk fieldwork studies should minimise the burden of the recruitment process, especially the consent procedures, which can be explicitly or implicitly connected to accountability. Standard Institutional Review Board (IRB) requirements for participant recruitment include: obtaining a signed informed consent from all participants; providing ample time (e.g., at least 24 hours) for potential participants to consider participation so as not to exert any pressure for rushed decisions; and providing only modest incentives for studies where no significant time commitment or risk are involved on the part of participants, to avoid coercing people into participation.

> **Tip:** Schedules can be set three or even four months ahead, which means that recruiting participants to research activities, like full day workshops, needs to be done in good time. They can also change at short notice which may mean patience and flexibility is needed for shorter appointments like interviews.

Written informed consent is the gold standard in healthcare research to protect patients and to make sure research protocols are understood and adhered to (see Chapter 1). This is most pertinent for clinical trials where care interventions are tested, which could have significant risks. In observational research, by way of a participant's signature, the consent form provides documentation of agreement to participate in a study. As such, the signed consent form provides protection to the researcher in cases when a participant may turn around and deny ever having consented to participation. It also provides evidence for who has participated in the study to deter fraud in research. Written consent may also promote a sense of formality where staff who have signed the consent form feel a stronger commitment to contribute to the research. The process of gaining informed consent can be an opportunity to explain and reassure people about the value of their participation. For example, if a study involved observations of a medical device, you should reassure participants that it is the device which is under study and not them. Also, if errors are observed you should explain who this information would be shared with, how, and under what circumstances.

In a study conducted by one of the authors, a potential nurse participant commented that "I don't mind participating, I just don't want to sign anything." This appears to suggest that written consent can be a barrier to participation. We believe this is mainly because of the clear paper trail that could implicate the participant if something went wrong. This reinforces the need to properly describe what information is gathered and under what conditions it would be shared with whom. More generally, informed consent can take a long time which many clinicians do not have. For low-risk qualitative studies with no material ethical issues, there can be alternatives, such as verbal or passive consent, if there is good reason for those (see Chapter 1).

Fieldwork researchers who are new to the healthcare context often do not realize that they have the option to present well-substantiated arguments to IRBs as to why other forms of consent, or waiving the requirement for consent, may be better suited to low-risk observational studies. For example, in one of the authors' studies, replacing the requirement for clinician-signed consent with that for verbal consent resulted in highly increased study participation rates. In another study, one of the book authors collected data during multidisciplinary team meetings. Getting signed consent from each individual before the start of the meeting would have been too disruptive to the clinical practice and therefore a strategy was adopted where the researchers emailed all attendees in advance and requested written consent only from the team leader, while the rest of the clinicians provided verbal consent at the start of the meeting. Hence, the upfront time cost in deciding on and justifying alternative recruitment methods is likely to pay off with improved recruitment rates.

In some intended research situations even verbal consent may present a major disruption and a consent waiver may be considered. However, a consent waiver is not standard practice and requires good justification that goes beyond anecdotal comments or clinician preferences.

Researchers should carefully consider recruitment strategies, the benefits and burdens associated with each strategy, and where appropriate, try to minimize aspects related to potential accountability and the amount of paperwork and time burden associated with the consent procedure. No specific justification is required by IRBs when using signed consent procedures, as this is expected.

For verbal consent and consent waivers, researchers need to justify the appropriateness, benefits, and risks to participants, and the necessity of the chosen method. This is described in detail by Sarcevic (Volume 1: Chapter 2).

> **Tip:** A project enabler can facilitate access and connections, particularly at the beginning of your study. A project champion can be involved in coordinating and promoting aspects of your project throughout your study.

> **Tip:** Carefully consider the pros and cons of each consent strategy within the context of your study. Signed consent is the gold standard, but alternatives, such as verbal consent or a consent waiver, may be appropriate for some studies and may improve recruitment.

3.4   HOW DO I MANAGE RELATIONSHIPS WITH PATIENTS AND CAREGIVERS?

Some research may involve not only clinicians, but also patients and their caregivers. Research involving patients and caregivers could occur in or outside the clinical setting (e.g., if the target population is self-managing a chronic condition, such as hypertension or diabetes, or using a home medical device as described in Volume 1: Chapter 11). For situated studies, fieldwork may have to be carried out in a person's hospital room, private home, or when the patient is on the go with mobile medical technologies. This type of research presents a range of complexities that a researcher will need to navigate, including getting to know and understand your participant population, over-coming recruitment challenges, and maintaining an empathetic relationship with patients.

Some of the advice on working with clinicians, such as gaining written informed consent, verbal consent, or a consent waiver, can apply to patients and carers too. The appropriateness of the consent strategy will depend on the ethical issues, risk, whether any patient data is collected, and whether the interaction with the patient is direct or merely on the periphery of a study. For example, what form of consent is appropriate if a researcher is shadowing a nurse to observe their practice on a medication round? Full written patient informed consent can be disruptive to the nurse's work, and disproportionate when the patient is not the subject of research and no patient information is gathered. Additionally, it has the potential to worry patients who do not want to read, understand, and sign forms if they are very unwell. If the patient is the subject of research and information is gathered from them directly, then written informed consent should be sought.

3.4.1   HOW DO I LEARN ABOUT TARGET PATIENT POPULATIONS BEFORE STARTING MY STUDY?

User research is an important first step in many domains of fieldwork, but particularly in the health-care domain as the researcher should be sensitive to participants' experiences around their health, which in turn can help in planning a study. In clinical settings, there may be opportunities to "be around" the patient population of interest, but this access can be awkward or in some cases not possible because of privacy concerns. In non-clinical settings, this type of user research can be even more difficult without a physical location where you could meet potential participants on a regular basis. Understanding the target patient or caregiver population can be achieved for both settings, albeit through less traditional methods than being present in the location. The following strategies can be applied for both settings:

- **Literature Review.** As a first step, a basic literature review should be undertaken to prepare the researcher. This review should include both scientific and non-scientific literature, as well as information from websites such as Wikipedia. The goal is to understand the patient condition, symptoms, implications for quality of life, and treatment. You should be able to get an understanding of what it is like to have this condition or care for someone with the condition. The literature review will also give an idea of the limitations of the target population; for instance, they might be too ill to conduct long interviews or collect diary information (Volume 1: Chapter 11).

- **Support Groups.** Attending support groups can be a way to start to understand the user population. Agreement should be sought from the organiser, and the group should have some indication of who you are, even though this would be prior to starting actual data collection. There are some patient populations that may not have support groups (e.g., migraine sufferers), but for most conditions there is a multitude of support groups that patients and caregivers attend (e.g., for diabetes, cancer, HIV, and addictions). This of course must be approached carefully: although many of these groups are very open to research projects, they can be very personal and private places for some.

- **Online Social Media.** Surveying online forums, patient blogs, and other social networks can also be used as a less invasive way of understanding and having access to the target population of patients or caregivers. There are communities on established social networking sites that cater to almost any patient or caregiver population, many of which can be surveyed without joining. In addition, there are other online forums and blogs that are also a rich source of information, and there may even be a connected community, such as The Diabetes Online Community (also known to its members as the "D-O-C").

- **Convenience Sampling.** Convenience sampling can be used to access members of the target population that you might already have access to. This can mean friends, family, acquaintances, people who are part of your institute, etc. Of course, this can produce its own hurdles, such as privacy concerns or your personal relationship influencing what they are willing to share, so this strategy is best used only for pilot studies or background data gathering.

- **Self-Studies.** Self-study is a method where the researcher uses a technology of interest over a period of time as though he or she is the patient or intended user. This method allows the researcher to become more familiar with some of the patient population concerns. For instance, one of the authors conducting research on Type 1 Diabetes mobile technologies used a wrist blood pressure monitor for three months in order to explore the experience of regularly using a medical device in everyday life (O'Kane et al., 2014). Although it should not be used in isolation, this method is an easy first step toward learning about the target population's experience of using technology in real life.

- **Accessing Clinicians.** Talking to clinicians about patients and their caregivers can also be a helpful way of understanding the target population of a technology study. Talking about specific patients and their conditions must be handled carefully though, as it may break patient confidentiality. It should be borne in mind that being a clinician is not the same as having the condition or being a caregiver. Nevertheless, clinicians know a lot about patients' and caregivers' experience, just from a different perspective. In addition, clinicians may provide an insight as to the technology choices recommended to patients. For instance, in the experience of one of the authors, a diabetes specialist nurse at a large hospital in the U.K. was the person who selected a glucose meter device for a newly diagnosed patient. The choice she made for patients had significant use implications because many patients do not change devices for years after this initial decision is made for them.

- **Patient and Public Involvement (PPI)** activities can help researchers learn about their target population and help with recruitment. For example, PPI representatives could review advertisements, and participant information sheets and give advice on recruitment strategies. PPI is discussed at length in Chapter 1 of this volume.

Tip: Taking the time to understand patients and caregivers is an important first step before starting the study. Your understanding of the patient condition and/or the caregiver challenges will inform the methods of recruitment, as well as your study protocol.

3.4.2   HOW DO I GAIN ACCESS TO PATIENTS INSIDE CLINICAL SETTINGS?

There are a few distinct options for accessing patients within a clinical setting. The researcher can approach patients themselves, but it is advisable to take guidance from staff about who it is appropriate to approach and talk to, e.g., some patients may be too ill. Alternatively, the patient recruitment process could be facilitated by a research contact at the hospital (e.g., the study champion or local coordinator) and conducted by the clinical care team for the target patient population. In this case, to help the clinical care team in the identification of potential participants, the researcher should provide clear, comprehensive inclusion and exclusion criteria, specifying which patients are eligible to take part in the proposed study. In some cases, the only inclusion criterion may be presence of the patient on the ward, or ability to communicate in the local language. In other cases, criteria may specify requirements such as age range, gender, specific diagnoses, etc. In addition, exclusion criteria must be defined, such as for example patients who are unable or unwilling to give informed consent, or who take certain medications that may interfere with the study targets. Using these criteria, the clinical team can then distribute information about the study, on behalf of the researcher, to potential participants. For this purpose, participant invitation letters and detailed information sheets including contact details of the researcher and information on how to get involved need to be provided to the clinical team. The researcher can then seek consent of patients wishing to take part in the study. It can sometimes be beneficial for the researcher to manage the recruitment alongside the clinical team in order to ensure that the inclusion or exclusion criteria are met satisfactorily. In studies conducted by one of the authors, leaving recruitment control to individuals outside the research investigator team led to inappropriate recruitment and tension between the researcher and the clinicians. Other scenarios include the clinical care team requesting consent from the patient too, but this is likely to only be for studies where there might be clinical choices and consequences that need to be discussed with a qualified medical professional.

3.4.3   HOW DO I GAIN ACCESS TO PATIENTS AND CAREGIVERS OUTSIDE CLINICAL SETTINGS?

Recruitment strategies to access patients and caregivers outside clinical settings can vary depending on the inclusion criteria and the target population. Accessing patients outside clinical settings in instances where they are recruited without using formal healthcare channels can sometimes mean that lengthy IRB-based ethics procedures are not necessary. In these situations, independent review from university- or institution-based ethics procedures may be adequate (see Chapter 1). However, recruiting patients outside clinical settings can be difficult as there is no standard way to access participants. Available channels for recruitment include online forums, social media, local support groups, non-profit organisations, pharmacies, and clinics. Sometimes charities and patient advocacy organisations will insist on review by an ethics board or committee within the healthcare system before being involved, while others will be happy to send out notices through their email lists. These different channels usually limit your reach to a specific cohort of the target population of patients and caregivers, as explained below. The limitation might be magnified if word of mouth is used in continuing recruitment—for instance, if a recruitment notice is shared on Facebook or Twitter—as it typically results in *snowball sampling* of friends and friends-of-friends within the same cohort.

Online recruitment through social media forums, such as Reddit, Craigslist, or Gumtree, can be a quick and easy way to reach out to a specific target population, but this may not always be successful depending on location, timing, and population. For instance, one author recruited patients on Reddit and found that this channel gave access to more technically savvy and younger participants, and garnered interest mainly from patients who were already participating in many research studies.

Recruiting from local support groups that meet in person is a good way to recruit face-to-face and be able to explain and answer questions about your research. This approach requires an understanding of who makes up these groups; for example, recruiting from diabetes support groups that meet once a month might mean that the participant pool consists of older adults who are unable to connect with peers online, as they are not comfortable using the Internet.

Recruiting from email lists, through recruitment companies, and not-for-profit organisations' social media pages (e.g., charities with Twitter and Facebook channels) can be used to access the target population of patients and caregivers. Each of these avenues has its own constraints; for instance, with university email lists you would be accessing young and well-educated patients, while going through not-for-profit organisations you might be appealing to people who are already very engaged in caring for their condition.

Posting physical posters and advertisements at pharmacies and grocery stores is another way to get access to the target population. Of course, this requires quite a lot of legwork and some stores may not be willing to advertise the study. In previous research, one author found that in different settings this worked with various degrees of success. For instance, it was very successful to post recruitment notices in smaller grocery stores in Cambridge, U.K., but the same method garnered no interest in larger grocery stores in Toronto, Canada, or Los Angeles, U.S.

Recruiting from a clinical setting, whether it is a hospital department or a local doctor's office, is a way to get a very specific user population. Unfortunately, it can also mean that these patients are people with more complications or having difficulties with self-management. Also, the geography of the physical institution might limit the range of demographics. For instance, recruiting in an institution located in a very affluent area of a city will predetermine the cohort. In addition, this typically means that ethics procedures for that health institution have to be followed (see Chapter 1).

> **Tip:** Understand different recruitment options and their potential impact on findings. Each option will impact the participant pool in a particular way, so sometimes a mixed-method sampling approach might be preferable. This is a good opportunity for "reflexivity" as described in Chapter 2.

3.4.4   HOW DO I MAINTAIN THE RELATIONSHIP DURING DATA COLLECTION?

Engagement can be difficult over time when there is no direct benefit to the participant. The researcher must be cognizant of the vulnerability of the participant and the possible risk that the project might compound their emotions related to their medical condition (as experienced with home haemodialysis studies in Volume 1: Chapter 11). Issues of rapport and managing the researcher-participant relationship are expanded upon in Chapter 2 of this volume (e.g., being reflexive about the impact of relationsips on research, becoming attached and detached from people in context, and not being exploitative).

The strategy of data collection must be well thought out by the researcher in order to maintain respectful relationships with patients and caregivers inside and outside clinical settings. Sometimes, situated observational studies may be difficult to achieve. For example, the study of mobile medical technologies such as glucose meters or healthcare apps may not allow a researcher to gain access to a patient at all times the devices are being used. In such cases, contextual interviews can be conducted in coffee shops or cafés that participants might frequent with their medical technology and feel comfortable in when discussing their technology use. However, these settings may introduce the risk of compromising personal health information by discussing it in a public space. Data gathering techniques and considerations are elaborated on in the next chapter.

3.5   SUMMARY

Managing relationships in any healthcare research agenda is important for access and to ensure the long-term success of the project. In this chapter, we have outlined some of the healthcare dynamics that might influence fieldwork studies, particularly with regard to the hierarchies inherent in health systems and the more subtle differences between healthcare settings. An awareness of the healthcare dynamics that could affect your study will help you throughout the research and ensure that you do not step on any toes. Additionally, tips were given to negotiate access to different healthcare settings. Access to the setting can be achieved through organised means or simply through interested stakeholders who can introduce you to the right people. Finding a key enabler who will facilitate your project is a very crucial step to setting up this access. Once your foot is in the door, it is important to develop and maintain relationships with your research subjects and others within the healthcare settings. Identifying a study champion among your target participant group is also critical, as this person will be your inside coordinator. Some of the strategies you should employ to establish interest in your study include being able to present your research concisely and simply, being there day to day, and establishing an identity among staff and becoming recognisable. Finally, this chapter also outlined some issues that may arise when conducting healthcare research that is patient-focused, such as research on home healthcare or mobile medical devices. Your study may require creative recruitment methods and you should invest time to understand your target population. This chapter laid out some of the social complexities and considerations when designing and conducting fieldwork in healthcare.

CHAPTER 4

Practicalities of Data Collection in Healthcare Fieldwork

Katherine Sellen, Aleksandra Sarcevic, Yunan Chen, Rebecca Randell,
Xiaomu Zhou, Deborah Chan, and Atish Rajkomar

When you begin your research project, data collection may not be the focus of your plan as you craft your study design using a theoretical or applied motivation for the work. Once the research begins, however, you will quickly discover that data collection becomes critical to study success. Developing approaches for data collection may even become the focus of the research study itself, as you confront data collection challenges in the healthcare context. Gathering clinical and other data in healthcare settings can present many practical and ethical challenges, as well as unexpected opportunities.

This chapter covers some themes that, in our experience, are specific to data collection in healthcare fieldwork. We present several commonly used types of data collection techniques and discuss specific issues that you might need to consider when planning or carrying out fieldwork. We also highlight and explore unusual and rich sources of data that may be easily overlooked. Finally, we look at issues of sampling and triangulation.

The table below lists and describes the content of each section.

How might different perspectives on data affect my study?	The way in which data is used and perceived may differ between healthcare research and HCI research, affecting study design and data access.
What data collection techniques will help me answer my research questions?	There are many different data collection techniques that might be useful for different types of research questions. Observations, interviews, focus groups, video recording, and device logs are reviewed.
What other sources of data should I be aware of?	To enhance your study, consider making use of clinical data, error reports, and other types of data available in the field.
How much data should I collect?	The amount of data available may be large so an effective sampling strategy is needed. Triangulation can increase the robustness of your study.

4.1   HOW MIGHT DIFFERENT PERSPECTIVES ON DATA AFFECT MY STUDY?

One aspect of data collection in healthcare fieldwork that you should familiarise yourself with at the outset of a project is the difference in tradition and perspectives on data between HCI research and research in healthcare. Considering the differences between these fields allows you to develop a data acquisition and analysis plan that can benefit both the non-medical- and medical-based aims of a project. This is an important consideration for effective and successful long-term collaborations that involve an interdisciplinary team. A good starting point is to consider what each field values in terms of research outcomes.

Typically, outcomes from a clinical perspective include those motivated by a rationalistic approach, which results in studies that provide data on clinical significance, effectiveness and efficiency, medical errors, performance-based metrics, workflows, barriers and facilitators, teamwork and decision making, and training. This approach fulfils the need for quantitative data that support decision making in practice and the development of evidence-based practice. Even when the goal is to understand the social aspects of technology use in practice, the perspective from which research is carried out is often tacitly rationalistic (Kaplan, 2001). For example, broad concepts such as adoption, impact on the patient, and impact on staff may be captured through measurements of the number of transactions over time, number of clinically significant errors, and the time it takes staff to complete a task. This approach is also common amongst human factors researchers.

In contrast, outcomes from a computer science and HCI perspective often reflect different values and concerns. In this case, the emphasis may be on qualitative research that can provide a variety of different types of knowledge, both for practitioners and researchers, at the theoretical and practical levels. Typical research goals might be to: provide design requirements or principles for a specific domain or user group; apply and explore new theoretical perspectives; explore the use of new technologies or applications; develop an in-depth understanding of user behaviours, values, and socio-technical considerations; develop usability techniques and methods; or develop or extend design methods. Given this varied set of possible research goals, there tends to be room for a wide range of methodological approaches and strong support for new ideas and techniques. Here, the goal is not to add to the evidence base in a rationalistic way but to explore new possibilities in theory and technology application that might inspire development by others.

Differences between the aims and outcomes of these two approaches to research, and consequently to data collection, can sometimes pose challenges for collaboration between researchers from HCI fields and researchers from medical fields. These challenges can be wide ranging and fall beyond data, such as choice of funding source, study design, use of the literature to inform design, timeframes, analytic techniques, and publishing and knowledge translation. When you consider data collection strategies, recognising these differences in aims and outcomes is the first step to effective collaboration. Quantitative and qualitative approaches can complement each other and so you may choose to use a combination of such approaches (triangulation is covered later in the chapter).

4.2   WHAT DATA COLLECTION TECHNIQUES WILL HELP ME ANSWER MY RESEARCH QUESTIONS?

The first step in planning data collection is to think about what data you need to collect in order to answer your research questions and how you might go about collecting this data. Different data collection strategies will be needed depending on whether you want to gather data on a particular practice or how a particular technology is used (typically gathered using observations) or on healthcare professionals' or patients' perceptions of that practice or technology (typically gathered through interviews or focus groups). If you want to understand users' specific interactions with a clinical system to identify design and workflow issues, you may need to look at software logs. If your research question is concerned with information or events that are measurable, such as task times, interruptions, and certain kinds of errors, you might consider a data collection approach that is more familiar to your healthcare collaborators. We have arranged our description of data collection techniques according to the types of data the techniques allow you to gather. It is important, however, to remember that for many research studies a range of different data collection techniques including both qualitative and quantitative approaches may be required (Volume 1: Chapter 1).

Tip: It may sound obvious but use your research questions to determine your data collection techniques. The research questions should drive the choice of technique and there should be a clear link between the questions and the technique.

Figure 4.1: Different methods of data gathering have different affordances and constraints, and you can't record everything! What will you choose to look at, record, and how will this impact your results?

4.2.1   WHAT DATA COLLECTION METHODS CAN I USE TO UNDERSTAND PRACTICES, INTERACTIONS, AND BEHAVIOURS?

If the goal of the research is to understand practices, including workflow, use of paper-based documentation and electronic health records (EHRs), communication and coordination among team members, or decision making, data could be collected using common ethnographic approaches such as observations, shadowing, and audio and video recording.

Observations

Observations are a popular method of data collection in healthcare settings. Observations usually follow in the ethnographic tradition, where the researcher spends a period of time in the setting of interest (LeCompte and Schensul, 1999). Observations in HCI were initially motivated by design and the recognition that if we want to develop technology that fits with the way people work, it is necessary to base those designs on a detailed understanding of the reality of the intended users' everyday work, rather than on idealised and simplified accounts that may be reported in interviews. In healthcare, observations have been used to inform the design of technology (e.g., Volume 1: Chapters 2, 5, and 7) and to understand how healthcare technologies are used in practice (Volume 1: Chapters 1 and 3).

The details of observations are typically captured in fieldnotes. You want a notebook for recording your fieldnotes, one that is small enough to not attract too much attention and can be easily slipped into your pocket. It is important that fieldnotes are written up as soon after observing as possible (Volume 1: Chapter 5). What you record in your fieldnotes will depend on the focus of your research. One of the benefits of observations is their flexibility. You may start your observations with a broadly defined research question and keep the scope of your fieldnotes wide. In this initial phase, it is important to observe at different times of day and different days of the week. You may find that working together with a healthcare professional, or someone with a healthcare background, helps you to understand details of medical practice that you might otherwise not appreciate. As you become more familiar with the setting, or in response to interesting themes that you had not anticipated before the observations began, you can start adapting your observations and your fieldnotes to focus on particular topics. For example, in her Ph.D. research, Randell (2003) began her observations in an intensive care unit (ICU) with a general interest in the role of technology in medical error. It was only when the ICU purchased a new haemofiltration device and the nurses began to develop strategies for setting up and using the device that the issue of customisation of medical devices was highlighted, drawing her attention to other examples of such customisation. Similarly, starting with general observations of nursing documentation practice, Zhou et al. (2009) gradually focused her inquiry on two particular working documents after the hospital adopted a Computerised Provider Order Entry (CPOE) system, which led to the disappearance of patients' psychosocial information in written format in nursing documentation practices. Both of these examples successfully combine two different modes of qualitative research: doing exploratory research to discover a question and doing directed research to find an answer.

While the researcher is primarily an observer, you may be able to move toward deeper embedding in the study context if you are able to spend time building rapport and trust with your participants and have senior-level support for your activities (Hammersley and Atkinson, 1995; see also Chapter 3, this volume). Groth and Frykholm (Volume 1: Chapter 7) were able to share the office and break room at the gastrosurgical department they were studying, allowing them to complement their more formal observations and interviews with informal conversation and overhearing of others' conversations, enabling them to better understand the culture of the department. Similarly, for her study of trauma resuscitation, Sarcevic (Volume 1: Chapter 2) had a dedicated desk in the office adjacent to the emergency department, shared with personnel involved in research and administrative aspects of trauma resuscitation. How you present yourself, including the very practical issue of what to wear, can also affect how staff perceive you and your ability to blend in (Volume 1: Chapters 1, 2, and 6; see also Chapter 2, this volume).

Shadowing

When starting your observations, you need to decide whether to shadow particular staff members, to find somewhere to sit or stand where you are not in the way but can still observe the activities of interest, or a combination of the two. The strategy you choose will depend on the setting, the focus of your research, and personal preference, and may change as the study progresses. Finding a convenient place to locate yourself can be challenging (see for example Volume 1: Chapters 2 and 3) so shadowing staff can often be the only way to gain access to the activities you are interested in, even though this approach presents its own issues (Volume 1: Chapter 5; Zhou et al., 2010). Try to vary who you shadow (e.g., staff with varying levels of experience) and, if relevant to your project, to shadow a variety of professional roles.

Video and Audio Recording

Video recordings can be an important source of data for healthcare fieldwork, allowing for detailed analysis of interactions with technology, and both the verbal and nonverbal elements of communication and collaboration (Hindmarsh and Pilnick, 2002; 2007). Video recordings are also a valuable source of data when working collaboratively because they allow for later detailed review and analysis amongst members of the research team (Vincent et al., 2004). There are, however, limited situations in which it is possible to record video data without compromising patient confidentiality while still gathering data that is valuable for the research. Video recording as a method of data collection is most feasible in the less public areas of hospitals, such as operating theatres (see for example, Mentis et al., 2012) and meeting rooms (Kane and Luz, 2006), and in clinic settings (Greatbatch et al., 1995; Als, 1997), where there is less movement of people in and out of the setting.

Choosing when to record video data will depend on your research question. If your research question involves emergency care, it may seem impractical to use video as emergency situations can happen at any time of day and night and may be infrequent. Similarly, if you are interested in changes over time, depending on your timescale, it may be impractical to maintain and capture video or audio data over a long time period. One way in which to manage the capture of video data for infrequent events and events unfolding over time is with motion sensing or RFID systems that are triggered by your participants' activities (see Taneva and Law, 2007; Sellen et al., 2010).

As a supporting technique, you may wish to consider running simulations in a controlled setting for specific situations, use of specific technology, or detailed task work. This can be a valuable standalone approach, but in the case of fieldwork, this technique can provide detailed interaction data in a controlled manner to complement *in situ* video recording. This might be particularly useful for studies addressing human error and detailed cognitive processes that are difficult to record *in situ*.

An important issue to consider when deciding how much video data to collect is the additional time that is required for analysis of video recordings. Typically a general guideline is that for every 1 minute of video recording 3 minutes of analysis time will be required so that, for example, a video recording of one 40-minute end-of-shift handover will require a minimum of 2 hours of analysis. Therefore the use of video recording should be balanced with the availability of time to make best use of the video during the analysis phase of research.

Where video recording is not feasible, you may instead choose to supplement your fieldnotes with audio recording. Audio recording can be useful when you wish to capture activities that focus on oral communication, such as handovers and multidisciplinary team meetings, allowing you to focus your attention and fieldnotes on the non-verbal activities, such as gestures, facial expressions, and the documents and artefacts that are referred to in the conversation (e.g., Volume 1: Chapter 5). You may also want to use audio recording to quickly dictate what you see, when you do not have time to write detailed fieldnotes.

Additionally, where you have permission, it is useful to take photographs, to capture the physical environment and artefacts of interest, such as medical devices and information sources that are used in support of work (Volume 1: Chapters 3 and 11). Where you do not have permission to take photographs, you can sketch the physical environment and artefacts of interest. It may also be possible to take examples or copies of work artefacts, such as handover sheets with any patient identifiable information blanked out or blank versions of such forms.

What Data Collection Methods Can I Use to Understand Perceptions, Ideas, and Relationships?

If the goal of your research is to elicit opinions, perceptions, or design ideas, or uncover relationships between different agents involved in healthcare work, you could collect data using interviews and focus groups.

Interviewing Healthcare Professionals in Clinical Settings

Researchers typically conduct interviews with healthcare professionals to elicit opinions and perceptions on the work and technology, improve their understanding of particular events in a complex setting, or supplement or validate observational data (e.g., Gurses et al., 2009; Rajkomar and Blandford, 2012; Sarcevic et al., 2012). Before undertaking interviews, whether with healthcare professionals or patients and their carers, it is necessary to develop an interview topic guide, although the content of this topic guide will depend on what the interviews seek to find out and will likely evolve as the research progresses (Arthur and Nazroo, 2012).

Scheduling interviews can be challenging as many clinicians are over-worked, with little time to spare for other activities. Interviews may be cancelled at the last minute due to pressures of work. Asking clinicians to be interviewed during a break time may be challenging as well: when they take a break, that means they need a rest and may not want to talk about work, and such timing of interviews may be looked upon unfavourably by ethics committees. The rest time may in fact be part of a unionised work contract and therefore not available for research purposes. The challenges of setting up interviews are elaborated upon by Furniss (Volume 1: Chapter 3), Rajkomar and Blandford (2012), and Sarcevic et al. (2012). One solution is to conduct informal and intermittent interviews while undertaking observations when clinicians have some moments to spare. Interviewing during work hours has an added advantage of being able to look at and discuss the artefacts that support the work practices in the actual work context. You may also find that some healthcare professionals have allocated time that they can use for research, particularly in university-affiliated hospitals, which can be used for scheduling interviews.

While audio recording may be possible during more formal interviews when participants give informed consent, it may not be possible during informal interviews. If you are not able to audio record the interview, try to jot down or remember key words or phrases that the interviewee uses, and then write up your notes in more detail as soon as possible.

There is no consensus regarding how many interviews are necessary to provide an adequate understanding of attitudes and experiences within a particular setting (Britten, 2006). It is dependent on the range of participants to be included, the purpose of the interviews, and whether other forms of data will be gathered. It is necessary to balance the desire for data saturation with the need to keep the data set to a manageable size. However, a sample size of three participants per group (i.e., clinical role, or stakeholder group) is generally regarded as an adequate starting point.

> **Tip:** Make use of quiet moments when observing to conduct informal interviews with staff, eliciting their opinions and clarifying details of events that have been observed.

Interviewing Patients

You may also be interested in gathering patients' opinions, perceptions, and attitudes. Interviewing patients about their use of medical technologies has certain challenges. Some challenges related to life-sustaining technologies include discussing difficult and emotional topics while not interrupting treatment (Volume 1: Chapter 11). Issues of power may arise when interviewing patients, if the characteristics of interviewee and researcher are reminiscent of a power imbalance in other social interactions, based on for example gender, disability, or age (Lewis, 2012). These issues may require you to use a more dialogical approach (or participants as co-researchers), where topics and flow of conversation are led by the participant rather than the researcher (Volume 1: Chapters 10 and 11). Further issues regarding biases and reflexivity are covered in Chapter 2 in this volume.

The critical incident technique (Flanagan, 1954) can be a useful tool to elicit patients' experiences in interacting with home-based medical technologies. Typically, a participant is asked to recount incidents they have had with technology. This technique is helpful because it provides a clear focus to the interview, which participants can understand (i.e., incidents they had with the technology). It also allows you to elicit actual facts (incidents) from participants, rather than just general opinions and impressions.

Interviewing a patient about health topics can stir emotions, so it is important to be sensitive to the participant's emotional states during the interview. Additionally, you should be prepared to welcome other people, such as family members, to participate in interviews. Inviting them to interviews is important because they are often involved in patient care and may regularly interact with the technology of interest. Also, if undertaking the interview in a patient's home, denying their presence may feel awkward.

Focus Groups

Focus groups are a useful qualitative research method for exploring perceptions, opinions, attitudes, and beliefs of a group toward an area of interest. Health services research has shown that focus groups can be valuable in healthcare for soliciting the views of patients, caregivers, and healthcare professionals when the subjective viewpoint is of interest (ICES, 1999). This method can yield rich data from healthcare professionals including personal experience with healthcare processes, insights into complex problems, or perceptions of a proposed intervention. It can also be used to explore patients' experiences with healthcare services or a specific illness or disease. The focus group structure also provides flexibility for the facilitator to explore unexpected themes or topics that are discovered during the discussion, a common occurrence when conducting research in the healthcare context.

One of the challenges with conducting a focus group is that it is impossible to predict how the personalities of the individuals in the group and the resulting dynamic will impact the direction of the discussion. Be prepared for the discussion to deviate from the intended plan and to modify your plans to adjust to the ebb and flow of the conversation. Moderating a focus group is a delicate process of balancing the strong personalities with those who are less vocal and hesitant to express their opinions to ensure all participants feel that they are heard, understood, and that their perspectives are valued. Be aware of the formal and informal hierarchies in healthcare roles that inevitably affect the dynamic of the discussion and participation, and the potential for bullying or punitive behaviours to affect your research (Volume 1: Chapter 1). You may choose to have a fellow researcher with you to assist in managing the session. The role of the moderator is pivotal to the nature and quality of data collected through focus groups (Sim, 1998) and therefore you may also consider attending training in focus group moderation or bringing in an experienced moderator. Depending on the study objectives, a homogeneous group may create a more comfortable environment for open discussion and would also capitalise on participants' shared experiences (ICES, 1999). Interdisciplinary focus groups provide the additional benefit of exploring cross-boundary relationships, thoughts, attitudes, activities (e.g., handoffs), frustrations, and priorities.

Focus groups can also be useful in the process of designing for healthcare by creating an opportunity for participatory design (Glushko, 2013; Kusunoki et al., 2014). Participants who understand the complexity of the healthcare context bring valuable perspectives for validating user requirements, workflows, and prioritising goals.

There may be practical challenges for patients and carers physically attending a focus group, which could be a barrier to using this method for certain populations. This may be alleviated by fully compensating participants for the cost of travel and any alternative care arrangements.

As with interviews, there is no consensus on the number of focus groups that should be conducted for a single study, although it is generally recommended that more than one focus group should be conducted (Sim, 1998). Advice on the number of participants for a focus group varies, with 4–8 participants (Kitzinger, 2006) and 8–12 participants (Sim, 1998) both being suggested as suitable numbers.

4.2.2   WHAT METHODS CAN I USE TO GATHER DETAILED DATA ON INTERACTIONS WITH TECHNOLOGY?

Gathering data on detailed interactions with medical tools and equipment through observations and interviews can be challenging. Seeing data on monitors and capturing keyboard strokes and interface interactions in real time may not be feasible due to the sensitive and critical nature of the technologies, as well as the complexity of the interactions themselves (Volume 1: Chapter 11). Data interpretation can also be challenging, requiring detailed understanding of a medical task, procedure, or terminology. You may find that interactions differ between the healthcare professionals, patients, or carers who use the system or device, due to different makes and models that have different operating procedures or simply because the users have been taught different procedures for setting up and using the device (Volume 1: Chapter 11).

Rajkomar et al. (Volume 1: Chapter 11) describe a range of strategies that they tried for studying patients' interactions with home haemodialysis technology. Video diaries proved not to be practical due to the stress and demands of dialysis treatment, and they were unable to access institutional data on accidents with such devices or to access a device on which to do bench tests. If you find yourself struggling to gather the detailed data you want on interactions with a particular technology, one option is to shift focus to a more accessible example that will provide similar interaction types but with fewer hurdles (see for example Volume 1: Chapter 3). Cognitive walkthroughs (Polson et al., 1992) with clinicians during off-hours provide an opportunity for in-depth exploration of a particular task or workflow where the specific interpretation can be provided in a less safety-critical scenario.

Software Log Data

When wanting to study the detail of interaction with a particular device or technology, analysis of software and system logs makes use of data that is already collected as part of the security and privacy requirements in the healthcare domain. For example, access logs in an EHR system automatically record who has accessed which record at what time (AHIMA, 2011). Although such data are originally recorded for privacy and security purposes, the system logs can also be used to uncover usage patterns, reveal users' information needs at the point of care, and identify usability issues with the current system (Chen and Cimino, 2003; Jalloh and Waitman, 2006; Sellen et al., 2010). These automatically recorded logs also facilitate understanding of user behaviours that are beyond direct observation (e.g., how clinicians interact with the system at home).

These access logs kept for security reasons have limitations, however, since they may not capture keystrokes and screen interactions. Provided you have approval to do so, you can either install stand-alone data recording software, or add research-enhancing logging features to clinical systems or equipment. For instance, Zheng et al. (2009) reported on clinicians' behaviours while using a homegrown EHR system that allowed real-time capture of comprehensive user interface interaction events. Keep in mind that you will need to budget some time to work with the IT department to document and test that the presence of additional logging software will not interfere with the running of the system or equipment you are studying.

4.3 WHAT OTHER SOURCES OF DATA SHOULD I BE AWARE OF?

In addition to the data collection strategies described above, there are several sources of supplementary data that can be useful for fieldwork in healthcare.

4.3.1   CLINICAL DATA

Access to clinical data can be key to the success of a field study, depending on the nature of the inquiry. You may wish to undertake a chart audit, recording specific items of data that are recorded in patients' records, or you may instead choose to focus on clinicians' informal temporary documents, such as handover sheets. For instance, in a study exploring the types of information doctors document versus orally communicate, Zhou et al. (2010) gained access to patients' records in the EHR to study the *gaps* between the written and oral communication practices. Similarly, to study how nurses pass patients' information across shifts and what they record in their working documents, Zhou et al. (2009) sought access to nursing working documents, including both workload related information and patients' psychosocial emotional needs.

4.3.2   UNEXPECTED DATA

While it is important to carefully plan your data collection, you should also be open to the value of unexpected data. In particular, when undertaking observations, and certainly in the early days of fieldwork, the scope of fieldnotes should be kept broad, going beyond capturing the work practice and technology use to also capturing what may initially seem tangential, but that may help in building a broader picture of the environment and culture (e.g., unrelated conversations, unexpected events, emergency cases, and how clinicians managed those). You should also respond to interesting unexpected cases when they do occur. For example, Furniss (Volume 1: Chapter 3) describes a device which was almost dismissed as being unproblematic and uninteresting suddenly coming into research focus one Saturday morning. Serendipitously, he was present while complaints about the device's alarm revealed that staff had issues controlling it. Also, access to data and settings can unexpectedly change, requiring you to be flexible and creative in order to continue the study. For instance, Hilligoss (Volume 1: Chapter 6) found that access to patient data was unexpectedly rescinded, leaving a data gap in his study. In such cases, you may look for other sources of data to replace access to clinical records. These sources can include error reports, order slips, notes, and help messages left on or around the equipment.

4.3.3   REMOTE DATA AND OTHER STRATEGIES

Because access to staff or patient time can be constrained by the reality of work patterns and working agreements (e.g., union agreements), as well as by the unpredictability of activity patterns, fieldwork capacity, and availability of the researcher, it is helpful to consider alternative sources of data. These sources can include journals and event cards that engage participants in recording ideas and events (Volume 1: Chapter 9). Providing a video camera, Smartphone, or other recording device can also be an effective way of "seeing the issue" through the eyes of the participant (Volume 1: Chapter 11). The use of recording functionality on mobile devices may be the quickest and richest way of engaging clinicians when journals are not appropriate (e.g., time constraints or carrying a diary around serving as a barrier).

4.4   HOW MUCH DATA SHOULD I COLLECT?

Whatever data collection method or combination of methods you decide to use, a general consideration is how much data to collect. It is tempting to collect as much data as possible, or add a variety of settings and experiences to your list of data collection sites and events. The risk with this approach is an increase in the time and resources needed. For instance, you might be interested in medical handovers and decide to observe different types of handovers in different settings, which would require a lot of chasing around and many hours of observation (Volume 1: Chapter 5). In contrast, you could choose to concentrate on one type of handover, e.g., nursing handover, or one context for handover, e.g., inpatient ward. The ways in which you can use the data, in terms of answering a research question, will be different in each case. You may also have to compromise on what you can say about the results. For instance, it may be difficult to generalise across types of handovers if your observations are specific to one context or group.

4.4.1   SAMPLING STRATEGIES

Sampling refers to the decisions about settings in which to collect data and, within those settings, on which people and events to collect data, and at which time points (Hammersley and Atkinson, 1995; Miles and Huberman, 1994). A key distinction when talking about sampling strategies is between probability samples and non-probability samples (Ritchie et al., 2012). With probability sampling, the aim is to produce a statistically representative sample, with cases within a given population chosen at random. In contrast, with non-probability sampling strategies, the samples are not intended to be statisically representative but participants, settings, or events are chosen because they have particular features or characteristics. Probability sampling is considered to be the most rigorous approach for quantitative research (and is the approach to sampling that your healthcare collaborators are most likely to be familiar with), but is largely seen as inappropriate for the qualitative data collection methods described in this chapter (see Sudman, 1976, for more on probability sampling).

The most robust approach to non-probability sampling is *purposive* sampling, where samples are chosen by the researcher, using their knowledge of the population and context, for the purpose of the research. For example, you may expect differences in weekend vs. weekday events, busy vs. non-busy times, or differences between groups of people, such as clinical specialty, or progression of disease, and so your sampling strategy should include the pertinent time periods and people. Hilligoss (Volume 1: Chapter 6) used a sampling strategy that ensured he could contrast perspectives between specialties and settings to provide a richer understanding of perceptions and work practices. If your research question has a theoretical emphasis, you may also consider *theoretical sampling*, an approach to sampling where the theory and insights gained from fieldwork lead the sampling strategy as the researcher selects samples (participants, sites, etc.) to further develop and test the emerging theory (Glaser and Strauss, 1967).

In most healthcare field studies, gaining access to participants, events, or sites can be challenging and recruitment may rely on collaborators who play the role of gatekeeper (Volume 1: Chapter 2; Volume 1: Chapter 10). In these cases, *convenience* and *snowball sampling* may be the only practicable options. Convenience sampling involves working with all available participants or settings, while snowball sampling involves participants introducing the researcher to further possible participants. These sampling approaches are often a good option for marginalised groups (Noy, 2008), especially in non-institutionalised settings, such as reaching specific patient groups in their homes. While you may not be able to generalise with convenience or snowball sampling, you may gain valuable insights into otherwise under-researched topics or hard to reach groups.

4.4.2   TRIANGULATION

Triangulation refers to the use of different methods and sources to modify, confirm, and refine your analysis and interpretation (Ritchie, 2012). While conducting fieldwork, it is often advisable to use more than one method of data collection. You may decide to do this anyway because you realise that one type of data collection technique is not sufficient to answer your research questions. For example, often we want to know both how something happens and staff or patients' perceptions of it, so observations are regularly supplemented by interviews. Triangulation is often suggested to be a key way of validating research findings, increasing the robustness of your research. However, while we recommend the use of multiple methods of data collection, it is important to be aware that this is a topic around which there is much debate. Different data collection methods vary in the types of data they generate, which is why we have organised the data collection methods described above according to the type of understanding they provide, and so they are unlikely to generate perfectly consistent evidence. Consequently, it has been argued that the value of triangulation lies in extending understanding, through the use of multiple perspectives. For example, if the work practice you observe is different from what is described by staff in interviews, that does not invalidate the findings from the interviews but should encourage you to reflect on why what is described is different from what has been observed.

> **Tip:** Use multiple methods of data collection and reflect on the differences in what each method reveals.

4.5   SUMMARY

There are many issues that you will need to consider when thinking about data collection in healthcare fieldwork. We hope that there is enough detail in this chapter to provide you with a starting point for developing a data collection strategy in your work. There are many topics we would like to include and other methods of data collection that we have not mentioned, but the ones discussed here are, in our experience, the most commonplace, or require a specific approach in the healthcare context.

CHAPTER   5

Healthcare Intervention Studies "In the Wild"

Mads Frost, Cecily Morrison, Daniel Wolstenholme, and Andy Dearden

The focus in this chapter is on research that involves introducing new technologies into healthcare. Research that develops new digital technologies or systems to enable, encourage, or support health can make a significant contribution to our society. To achieve this aim, it is necessary to carry out appropriate evaluations in realistic settings, or "in the wild." This ensures the clinical practicality and efficacy of the technology and is necessary for it to be accepted into healthcare settings. There are many challenges, however, that can deter researchers from testing the clinical validity of their ideas, from the complications of multidisciplinary team working to ethics committee headaches (see Chapter 1). This chapter provides guidance to make it easier for HCI researchers to take the plunge.

Evaluations of new technologies "in the wild" are more formally referred to in healthcare settings as interventional studies. Interventional studies are ones in which a change is made to the way healthcare is delivered, such as with a new technology or clinical protocol, and the effects of that change are evaluated. These types of studies can be contrasted with studies that focus on how healthcare is being practiced currently without the introduction of any changes, e.g., Chapter 4 focuses mainly on these studies.

In this chapter we address some of the most common and frustrating issues that arise when carrying out interventional studies in healthcare. This chapter covers four aspects of interventional study design that have unique characteristics in healthcare settings: choosing a study design; behaving ethically and navigating governance procedures; choosing and accessing a healthcare setting; and building and deploying technology. Some of these themes have been discussed in previous chapters, e.g., Chapter 1 on ethics, Chapter 3 on accessing and understanding the healthcare setting, and Chapter 4 on study design. This chapter extends and contextualises this material for studies that involve the development and introduction of a technology.

The table below names and describes the content of each section.

What should I think about when designing a study?	Articulates ways to focus your research aims and match an appropriate study design to them.
How do I navigate ethics and governance procedures for intervention studies?	Describes research ethics challenges you may encounter and introduces a vocabulary that will help you manage them.
How do I choose and access a healthcare setting?	Differentiates between different healthcare setting choices and discusses the practicalities of access.
What should I consider when building and deploying technology in a healthcare setting?	Provides a *"real-world checklist"* to support the successful deployment of technologies in healthcare settings.

5.1 WHAT SHOULD I THINK ABOUT WHEN DESIGNING A STUDY?

When selecting a study design, the first step is to define, and subsequently refine, the aims of your research. The next step is to consider the level of evidence required given the maturity of the idea. Finally, there are other issues to consider, such as whether the technology should be evaluated stand-alone or within the context of a particular approach to service delivery. This section deals with these three crucial steps in choosing an appropriate study design.

5.1.1 HOW DO I DEFINE AND REFINE MY RESEARCH AIMS?

Interventional studies that evaluate a technology are designed to create a specific piece of knowledge (or, from a clinical standpoint, evidence). The research aims and questions of a study capture precisely what that knowledge will be. A research aim articulates the broader concepts a study will address. A research question defines the measure(s) that will be used. For example, "Understand the level of engagement with an online intervention" is a research aim while, "What percentage of participants are still using the online intervention at the end of the study?" or "What is the effectiveness of the intervention in maintaining adherence to a medication regimen?" is a research question.

It is particularly important to iteratively refine the nuance of the research aims before designing a study if working as part of an interdisciplinary team. Most health technology projects include technologists, HCI specialists, and healthcare professionals, each with different goals and knowledge requirements as captured by Sarcevic (Volume 1: Chapter 2). This can create at best, misunderstanding, and at worst, conflict. It is best to encourage negotiation of the research aims, rather than arguing over particular study design decisions. This will help maintain the coherence of the study design.

5.1.2   WHAT LEVEL OF EVIDENCE IS APPROPRIATE TO YOUR IDEA?

Once you have articulated the aims of your study, it is important to think about what level of evidence is appropriate to your idea. A new idea may require a small initial study to determine whether to pursue further, while an idea that has gone through several increasingly larger evaluations, may be appropriate for a substantial, long-term study. The U.K. Medical Research Council's recommendations for the design and evaluation of complex health interventions articulate five stages of evaluation organised along an axis of increasingly robust clinical evidence (Campbell et al., 2000). The complexity and cost of each study design also increases as you move through the stages. The model draws attention to the iterative nature of evaluation and helps researchers think about what level of investment in evidence is needed given the maturity of the idea or technology. These five stages also have correlates in HCI research. These are captured in Table 5.1.

Table 5.1: Clinical testing stages and their HCI equivalents

Clinical Testing Stage	HCI evaluation method
Theory	Conceptual test • Test low-fidelity prototype with target users • Qualitative study of current practice
Modelling (Feasibility)	Small test of prototype with target users in context • Test whether an intervention is used • Check for usability problems • Check for alignment problems with treatment or service
Exploratory Trial	Study with defined outcome measures • Pre/post studies
Randomised Controlled Trial	Controlled study in an appropriate context
Longitudinal Study	Longitudinal study • Study of large scale log-data, over months/years, from an established functioning service

5.1.3   WHAT OTHER ISSUES SHOULD I CONSIDER?

It is important to reflect upon the external validity and generalisability of findings given the study design choices made. There are a number of technology-specific issues to consider. Should the technology be evaluated as a stand-alone entity or as part of a specific service-model? While the former would be ideal, so we understand the strengths and weaknesses of a technology detached from any variations in service, it may not be practical. For example, it is difficult to evaluate an online intervention without assessing the service model that encourages its use. One should also consider the rate of change of the technology and the speed with which evaluations can be carried out to ensure the study is not out-of-date before it has started.

Iteration, an important part of technology development, is unfamiliar to many healthcare professionals. Iteration is difficult to include in study designs that may be acceptable to healthcare professionals. It is important to discuss iteration with research collaborators early in planning a study and consider what might be done as a pre-study. Remember that an interdisciplinary team can be divided by vocabulary. A legitimate HCI study may not be considered to be a *research* study at all by clinical colleagues. There is often space for both if one does not insist on calling both "research."

Theory is another important issue. Not all HCI research builds on theory (such as behavioural theories). However, interventions that draw on theory are easier to evaluate as they direct what should be measured (Klasnja and Pratt, 2012). The intervention itself is likely to be more effective as well (Webb et al., 2010). Chapter 4, in this volume, draws distinction between HCI and human factors research that is more interested in rich understanding and theory, and medical research that is more interested in outcomes. When it works well theory can be used and developed, while outcomes are positively affected.

> **Tip:** Consider what the primary research aims are, and how they reflect the stated interests and perspectives of patients, healthcare professionals, and colleagues from other disciplines.

> **Tip:** Consider how each aspect of your study design addresses the research aims while taking into account issues of access and ethics.

5.2   HOW DO I NAVIGATE ETHICS AND GOVERNANCE PROCEDURES FOR INTERVENTION STUDIES?

As discussed in Chapter 1, the issue of research ethics can be a source of considerable challenge for healthcare fieldwork; this is particularly the case when dealing with intervention studies. In our experience, the challenge comes not from any difficulty that HCI or interaction design researchers face in behaving ethically, but in complexities that arise in navigating the specific processes for the ethical governance of research in healthcare.

5.2.1   HOW DO I HANDLE ITERATION AND ETHICS?

One area that is problematic for healthcare fieldwork is the tension between faster iterative design and evaluation cycles and the more lengthy research ethics review cycles. This is a challenge because the processes and timeframes are at odds.

A key principle in medical research is that it is only acceptable to involve humans in research that is soundly designed so as to generate useful knowledge. If the research will not generate useful knowledge, then it has no benefit and so cannot justify any costs at all (in time and effort) to the human participants. For these reasons, many ethical review procedures expect that detailed research plans are specified and independently peer-reviewed before research is commenced (see Chapter 1 in this volume for further detail). Any changes in the plan require further review, which can be a lengthy process. This may be difficult if the intention of the research is precisely to develop and modify proposals in a flexible and iterative manner with shorter time-scales. A similar tension has been noted for Action Research in health. Khanlou and Peter (2005) suggest a process whereby each cycle of action research should return to the ethics committee for re-examination. However, this solution is not the only way that interaction designers and healthcare researchers can successfully collaborate in research. By understanding the specific terminology that is applied in healthcare and in healthcare research, interaction designers may find other ways to navigate the governance procedures in this domain.

5.2.2   WHAT HEALTHCARE RESEARCH TERMINOLOGY SHOULD I KNOW?

Useful terms for the researcher wishing to navigate this space are: *translational gaps*, *pre-protocol research*, *governance*, *service review*, *service audit*, and *service improvement*. These relate to terms for ethics and governance in Chapter 1.

A key concern in healthcare is the challenge of getting innovative and successful research ideas worked out in the day-to-day practice of healthcare. Healthcare researchers and practitioners often refer to this as *translation*, and there is a deep concern about how *translational gaps* can be addressed. Intervention studies using iterative research can be positioned within this discourse as an effort to smooth translation of ideas from "bench" to "bedside." In the medical tradition, *bench science* is typified by scientists in laboratories looking for a potential substance that produces a response in a test tube that might have a therapeutic effect in people. Bench science is naturally iterative, and is accepted as the domain of university labs covered by internal university ethical governance procedures.

Other early research that is typically governed by university ethical review systems is referred to as *pre-protocol* research. Chapter 1, this volume, discusses when ethical review is required from within the healthcare system. Pre-protocol work might be preliminary qualitative work to determine the feasibility of a potential research project, or gathering data to support an application for funding. Working with collaborators to develop designs may fall into this category, for example in Groth and Frykholm's (Volume 1: Chapter 7) study, the initial project meetings, prototyping, and evaluation of prototypes are not regarded as "research" requiring ethical approval from the hospital (though in some countries it might require approval through a university's ethical governance procedure). Similarly, Randell's (Volume 1: Chapter 5) discussion of establishing consultations and "patient panels" to guide the development of a research protocol would fall into this category.

As discussed in Chapter 1, *research ethics* is just one aspect of the broader category of governance. In the U.K., bench science and pre-protocol work can take place in a healthcare setting without requiring the full process of NHS research *ethics review*, although the work will be subject to clinical *governance*. Some forms of design research and product development work clearly sit within this category, and it is entirely appropriate that the research elements of the work, which typically lie in disciplines such as computer science, HCI, design, engineering, etc., are governed by university ethics procedures. However, this does not mean that there are no healthcare governance issues.

All activities in healthcare settings are subject to *governance* even if they are not subject to research ethics oversight. Governance procedures clarify who is taking decisions, who is permitted to perform what activities, and who is accepting responsibility for these activities. Chapter 1 expands on ethics and governance issues further.

An important distinction to note in healthcare is that research is usually defined as being activities above and beyond usual practice, which therefore require additional ethical review and governance. Activities that may resemble research but whose intention is to improve the delivery of care in a specific context, to test out interventions that have already been researched elsewhere, or to audit against standards tend to be classed as *service review* in the NHS (see Brain et al., 2011). Service review is subject to *clinical governance* that will have its own procedure in each organisation (perhaps requiring the approval of an audit or clinical effectiveness department), but this is distinct from the *research ethics review*. Other related terms are *service audit* as well as *quality and service improvement*. Projects working with patients and staff to explore innovations to improve a particular service might fall under these categories, rather than "medical research." The governance procedures for such work are typically more accommodating for iterative design research and other forms of interventionist healthcare research activity.

It is important, however, to be clear about whether or not a technology or service proposal has a direct role in the delivery of medical care. If so, then the technology may be classed as a *medical device*, a classification used for tools such as infusion pumps, X-Ray machines, robotic surgical equipment, etc., which would then (not surprisingly) be subject to extensive and very rigorous regulatory oversight in most countries.

Regulations are enacted at a national level. While there are many commonalities across jurisdictions, there are also important differences, so it is essential to become familiar with the relevant regulations for each intended market. Within North America and Europe, there are similar definitions for classifying technology: a product is classed as a medical device if it is hardware/software intended by its manufacturer to be used for diagnostic or therapeutic purposes. On this basis, many familiar items such as plasters (bandages) and thermometers (for recording body temperature) are classed as medical devices, whereas much health-related software (e.g., mobile apps to encourage behaviour change by tracking diet and exercise) is generally not, as it is not intended for clinical diagnosis or care. Medical devices are classed according to the risk they pose; for example, a thermometer is considered less risky than a pacemaker.

Low-risk devices are typically subject to very lightweight oversight, whereas high-risk devices are subject to rigorous approval processes administered by the relevant authority (e.g., U.S. FDA, Health Canada, Notified Bodies reporting to national authorities in Europe, etc.). For most devices, this will include conducting a clinical trial to prove it works as intended. There are many other critical things in the approval process to bring a device to market that are too numerous to adequately cover here. Once a product is on the market, there are further requirements for post-market surveillance to identify failures or possible harm, so as to ensure that products are as safe as possible (commensurate with their benefits). One standard that is more relevant for this audience than others is Standard ISO/IEC 62366, which covers the application of usability engineering to medical devices. Standard ISO/IEC 62366 includes the device's development and approval process, and post-market surveillance.

By being aware of different categories of device and using appropriate language when describing planned activities, researchers may find that their healthcare system can provide *governance* procedures that are better adapted to the needs of the particular study. Thieme et al.'s (Volume 1: Chapter 9) discussion of the importance of ensuring that their Spheres of Wellbeing were recognised as not being a *medical device* reflects this concern.

> **Tip:** Not all intervention studies will be defined as "healthcare research" as the healthcare system understands it. For example, it could be a service improvement or pre-protocol that could be subject to university ethics approval. Carefully consider how your work is defined and seek advice if unsure.

> **Tip:** At different stages of design and development consider what the accepted form of healthcare governance is for this stage, of this study, in this particular setting.

5.3 HOW DO I CHOOSE AND ACCESS A HEALTHCARE SETTING?

When carrying out HCI research on healthcare implementations, some time should be spent considering the most appropriate setting to evaluate the implementation. Attention to a strategy for gaining an appropriate level of access is also crucial to a project's success. This section considers these two parts of the research process.

5.3.1   HOW DO I CHOOSE A RESEARCH SETTING?

One of the main goals of HCI research in healthcare is to explore, introduce, or evaluate new ways of caring for people utilising technology support. Using novel technologies is usually intended to change the way that existing services operate, presenting challenges for how to involve stakeholders in their design, and then how to evaluate them. The first step in planning a study is to consider the trade-offs inherent in choosing a particular healthcare setting. Three broad approaches should be deliberated: (1) creating a new service around the intervention by introducing a new infrastructure and system, (2) build upon an existing service, introducing a new intervention while not formally changing the original service, or (3) formally changing an existing service to accommodate the intervention better.

Let us look at the trade-offs of each of these options with the example of online interventions for common mental health problems. One benefit of these types of interventions is that they reduce the stigma of care through enabling anonymous online contact. Their value would be best captured if a new service was created that allowed people to sign up to the intervention online and receive a telephone assessment. This approach, however, would require substantial financial commitment and time-consuming attention to logistics. Alternatively, working with an existing service requires convincing patients who were initially offered face-to-face treatment to use an online intervention. This contrast is likely to be off-putting to many patients and does not capture the particular population that may be keen on anonymous interaction. The trade-offs that you may like to consider in designing your own study are in Table 5.2.

Table 5.2: Trade-offs to consider when choosing a healthcare setting		
Setting	**Pros**	**Cons**
Creating a new service: creates new infrastructure and systems	Technology can be tested as proposed rather than compromise with existing systems	Difficult and expensive to attend to the logistics of creating a new service
Work with an existing service: extends upon existing infrastructure and systems as is	Evaluated in an existing setting likely to adopt the technology. Opportunity to learn from adapting to current practice	Challenge of fitting to current practice. If the existing service provides obstacles and barriers it could hinder demonstrating the benefits of the new technology
Change an existing service: changes existing infrastructure and systems	Easier than creating a new service. May showcase changes to services and the benefits of the new technology better	Substantial collaboration and buy-in is required from senior healthcare professionals

5.3.2   HOW CAN I NEGOTIATE ACCESS?

Once you have chosen a setting, the next step is to negotiate access. This is an extended and time-consuming process that goes far beyond just getting in the door. For many people, initial access is established through an existing partnership that has been created through a research grant or clinical champions, as discussed in Chapter 2. These perspectives on access to research sites and project management are captured well by Sarcevic (Volume 1: Chapter 2). In this section, we discuss those negotiations that need to take place following the initial granting of access, which are pertinent to intervention studies.

It is essential to discuss the level of technology development and support early on and manage expectations. Healthcare professionals are accustomed to products, not research prototypes. Products should have no bugs, are supported by 24-hour helplines, and are integrated into existing systems. This is not possible for research prototypes. While the level of technology development may need to be greater than for most HCI evaluations, it is essential to be clear and negotiate. For example, you could offer a dedicated person to address problems or questions with the system, but not integrate all functionality with existing systems. It is best to focus on those aspects of system development that improve the experience of use.

It is also important to take the time to understand the clinical and service processes in detail (e.g., assessment procedures) to ensure the study design is appropriate. It can be challenging to fully expose day-to-day operations that may affect the running of the study. For example, in one study (Morrison et al., in press) we were aware that our clinical partners were triaging patients through a questionnaire. We were later surprised to find out that those who did not supply an email address were automatically filtered out from our study when this was not part of our exclusion criteria. Encourage your collaborating healthcare professionals to map out relevant processes on paper and then talk through some recent cases. Request any related documentation. This approach will help to elicit the nuances of the process.

Finally, plan for service evolution between your initial discussion of a study through to its projected end time. Services are frequently reconfigured. A risk assessment should be done as to how the evaluation would be affected if there was a significant restructuring of the organisation, the clinical champion were to leave, or a change in national priorities occurred. Not least, good documentation can provide clarity of study purpose and commitments, ensuring efforts are not derailed by changing personnel. Meeting minutes and protocols are two common communication instruments for capturing what has already been agreed about the study.

> **Tip:** Intervention can be complicated. Make sure you have negotiated the level of technology development required for evaluation, understood the detail of the current process you will integrate the technology into, and planned for organisational changes.

5.4   WHAT SHOULD I CONSIDER WHEN BUILDING AND DEPLOYING TECHNOLOGY IN HEALTHCARE?

Once the relationship with a healthcare setting has been clarified, much work is needed to embed the technology in it before the study starts. Attention must be paid to both functional and non-functional requirements of the technology. In this section, we first present a checklist to prompt consideration of both types of requirements. We then look in detail at how to create staff and patient buy-in and support for an ongoing study.

5.4.1   WHAT FUNCTIONAL AND NON-FUNCTIONAL CHALLENGES SHOULD I BE AWARE OF?

When deploying healthcare technology "in the wild," there are a range of issues—some obvious and others more subtle, but still crucial. The more obvious issues for HCI researchers relate to getting the technology to work as technically intended. However, it is not enough to build something that works. Effort needs to be put toward successful deployment in your chosen healthcare setting to get results from an evaluation study. These two issues are captured by the terms functional and non-functional requirements. Functional requirements refer to those related to getting the hardware and software to do as intended. Non-functional requirements include activities related to a successful deployment, such as training and creating staff and patient buy-in.

Both functional and non-functional aspects of a technology must be considered to run a successful study.

Examples of these challenges can be found in the work carried out in the MONARCA project, a monitoring system for bipolar disorder patients discussed in Chapter 8, Volume 1. These entail everything from managing the support of the system and the infrastructure, to creating user guides for patients and clinicians.

The real-world checklist for developing healthcare-based systems, in Table 5.3 and 5.4, proposes questions to consider during real-world deployment in pervasive healthcare. The lists are based on work done by Hansen et al. (2006). While not exhaustive, they serve as a good starting point to prompt consideration of different aspects of technology development needed in these types of projects.

	Table 5.3: Real-world checklist for developing healthcare based systems—functional aspects
Equipment security	Is the equipment secure? Is there a risk of theft?
Environment	Does the environment pose special requirements on the equipment? Is the system going to be used outdoors? Can it handle vandalism? Can it withstand being dropped or cleaned?
Power	Does the system require a power plug? How long can it run without being recharged? How do you recharge the system?
Network	How will the device communicate? Does it require an Ethernet connection? Is the wireless infrastructure in place? Do you need to transmit data in an external network?
Space	How much physical space does the system use? Is there space on the wall for large wall displays? Is there table space for another computer? Do the doctors have enough room in their pockets for another device? Is there space on the dashboard for another display?
Safety issues	Will a system malfunction affect safety issues? What is the contingency plan in case of a full system crash? Will the system interfere with other systems? Can the system pose a threat to the user?
Debugging	If the system malfunctions, how do you find the error? Does the system store debugging information? How do you detect serious errors in the system? How is logging done?
System security	Does the system need to be secure? How does it keep information confidential and secure? Is there a concrete security risk?
Integration	Is the deployed system stand-alone? Does it need to communicate with other deployed systems or integrate with third-party systems? Is there a public API and converters for communicating between systems?
Performance and scalability	How does the system perform? Is system performance acceptable in the real-world setting? How many devices are needed for deployment? Does the system scale?
Fault tolerance	What happens when an error occurs? Can the system recover automatically? Can the daily system users bring the system back up to a running state? Is the developer team notified about errors? Is the system configured for remote support?
Heterogeneity	Does the system run in a heterogeneous environment? Do heterogeneous elements need to communicate?

Table 5.4: Real-world checklist for developing healthcare based systems—non-functional aspects	
Cost	What will implementation cost? Will scaling up the system affect the price? Is special equipment needed?
Deployment and updates	Does the deployment mechanism scale to a large number of devices? Can you update the system? How do you update the different devices? How is the software installed and by whom (e.g., by a nurse, a technician)? Are the devices accessible after deployment?
Usability	Will end users use the system? If so, how many? Can the average user use the system? Does the interface pose problems? Does the system's overall usability match the average user?
Users	Does it need to display content in a different format based on who is using it, e.g., nurses, doctors, or patients?
Location	Where will the targeted healthcare setting be? Ambulatory, home and mobile, emergency ward, or mental health? Will it be migrating?
Training	How do the users learn to use the system? Is it individual instruction or group lessons? Does the system need super-users? Is a manual or help function needed? How does the user get support?
Politics	Who controls the system? Does the system change the power balance in the user setting? Who benefits from the system? Is the person that benefits from the system the same as the person that provides data to the system? Does the system require extra work from users?
Privacy	Does the system reveal private information? What kind of personal information does the system distribute and to whom?
Adaptation	Is the organisation ready for the system? Is there organisational resistance? Will the system change formal or informal structures in the organisation?
Trust	Does the user trust the system? Is the information given to users reliable? Who sends the information?
Support	Will the developers support the system? Does the support organisation have remote access to the system? Will there be a hotline? Are there user guides for the participants?

5.4.2   HOW CAN I GENERATE STAFF AND PATIENT BUY-IN?

When dealing with non-functional challenges, staff buy-in is critical to the success of the study. Healthcare professionals' attitudes toward an intervention influence how they present and support it. It is important to emphasise the benefits of the technology to staff for both themselves and patients, citing any evidence that may be available. However, remember that healthcare professionals are unlikely to accept a technology that usurps their role, so careful phrasing is needed.

Staff training is also crucial to success. Demonstrations followed by several dry runs are likely to work well. Negotiating training time can be difficult for busy healthcare professionals, so ensure that training issues are discussed early and built into your project plan.

Not least, attention should be paid to how a new technology is presented to patients and the wording of any materials should be refined. It can be difficult for patients to understand how a new technology will help them. It is important to relate it to something that they know and that matches their goals and beliefs. For example, an online intervention can also be described as self-help conveniently accessed online. Patients in one of our studies preferred such a description, as it matched their image of themselves. Wordings should be tested and standardised (with some flexibility) before a new technology is introduced. Chapter 1, in this volume, discusses different mechanisms for patient and public involvement (PPI) to help get this sort of input and feedback.

5.4.3   HOW SHOULD I SUPPORT AN ONGOING STUDY?

As the study runs, regular communication of happenings and results helps to keep people enthusiastic about the study and the ideas. It also provides a channel by which staff and patients can raise any issues or discrepancies between their practice and how the technology works. This can be done through a mix of "push" and "pull" information technologies such as email or intranet websites, but even better if communicated in person, e.g., at ward meetings, where the issues or discrepancies can be handled directly. For busy ward staff whose work is not primarily computer based, remember that paper newsletters and posters can also help to keep stakeholders informed.

> **Tip:** Staff and patient buy-in can be essential for an intervention study, so think carefully about how to generate and maintain it.

> **Tip:** Different functional and non-functional requirements must be accounted for when preparing for an intervention study. Make sure these are considered and use the checklists in Table 5.3 and 5.4 to help.

5.4.4   HOW SHOULD I WITHDRAW FROM A STUDY?

Thought should be given very early on in any intervention study about what will happen when the study has finished, both to manage expectations and also to put in place arrangements for supporting or removing the intervention at the end of the study. Unless long-term maintenance and integration of the intervention was planned for from an early stage, it may not be possible to commit resources to maintaining the intervention and upgrading it as other systems evolve. If the intervention was highly regarded, it may be possible to find further resources to maintain it in service. Where this is possible, this is clearly an important way of delivering impact from the research, a topic to which we return in Chapter 6.

Where this is not possible, arrangements need to be put in place for smooth withdrawal of the intervention. As noted above, clinicians often do not distinguish between research prototypes and fully functioning systems, so they may naturally assume that the intervention will continue to be available after the study has been completed. They may feel let down if the researchers remove the intervention, particularly if it was perceived as being successful. In this case, it is important to manage expectations by ensuring that clinicians are aware of the need to withdraw it from an early stage of planning.

Figure 5.1: Post-study planning for intervention studies. Careful thought needs to be given to the end of the study before it begins. Make sure participants have clear expectations about what will happen so they don't feel let down.

Hopefully, in the longer term, these novel interventions will come to market and benefit patients and clinicians. It can be useful to communicate this longer-term picture, and the importance of testing before the product is released, to participants who help at different phases of the development cycle.

5.5   SUMMARY

Designing interventional studies is challenging, but the impact can be substantial. In this chapter, we have raised some of the issues that are specific to evaluating new technologies in healthcare settings. As you design your study, you will inevitably find that these sections are interdependent and that the resources available to you will shape many decisions. Below are a summary of issues that, when considered, should lead to a stronger study design and smoother running of the study.

- Gather and develop comprehensive stakeholder requirements and insights into current practice.

- Fully explore your options for study designs and how they relate to the level of evidence required and the resources available.

- Push for an iterative development process where appropriate.

- Select or develop technologies that are flexible and can be expanded so rebuilds are less likely in the face of new requirements.

- Consider the *"Real-world checklist"* to try and flush out problems that you might encounter in later stages.

- Be proactive in finding out the governance procedures that are required.

The more thought you put into these issues before you start, the smoother the study process will be. Although negotiating the hurdles discussed in this chapter can be tedious, it can lead to a very rewarding outcome—the contribution of innovative and well-designed healthcare technologies to our society.

CHAPTER 6

Impact of Fieldwork in Healthcare: Understanding Impact on Researchers, Research, Practice, and Beyond

Helena Mentis, Svetlena Taneva, Ann Blandford, Dominic Furniss,
Raj Ratwani, Rebecca Randell, and Anjum Chagpar

This chapter outlines various forms of impact your research can have as well as the steps you can take in determining and ensuring impact. Our motivation in this chapter is to discuss the reasons you should consider impact at the outset of your research and the possible paths toward having an impact with fieldwork research in healthcare.

Although field studies have contributed much empirical and conceptual knowledge on, for example, coordination, sense-making, usability errors, technology acceptance, and information seeking in healthcare settings (see Bardram et al., 2006; Mentis et al., 2013; Reddy and Dourish, 2002; Sarcevic et al., 2012; Scupelli et al., 2010; Tang and Carpendale, 2007; and Randell et al., 2011; as well as some of the case studies in Volume 1), a critical step to improving healthcare is to transfer this knowledge into a form that leads to improved health system quality and safety. This was referred to as the *translational gap* in Chapter 5. As researchers, conceiving of the practical implications of one's work and, more importantly, conveying and transferring those implications can be challenging. These challenges can be due to a number of factors; often the biggest obstacle is limited understanding of what forms of impact are feasible and practical, given the state of clinical practice, health policy, or existing system implementation.

In the following sections we outline various forms of impact you can achieve with your research. In Chapter 5 we discussed system design and implementation, which can be conducive for particular forms of impact. In this chapter we are focusing more generally on types of impact your research can entail. We then lay out strategies you can take in planning for impact from the beginning and following through to achieve it.

The table below names and describes the questions and concerns of each section.

Why is impact important?	Impact is important because as researchers we should strive to improve healthcare quality, safety, and/or processes with the work we do.
When should we consider impact?	Impact should be considered at the start of a project where pathways to impact should be planned. Various steps are discussed for considering impact.
What types of impact are feasible?	Six potential areas of impact are outlined: researchers and practitioners; scientific knowledge; technology design; practice; society; and economy.
How can we transfer the findings of fieldwork to achieve impact?	To bridge the translational gap, it is necessary to write for different audiences and engage with decision makers.

6.1   WHY IS IMPACT IMPORTANT?

To begin with, what do we mean by "impact"? Many of us who choose to conduct research in the health domain have the underlying motivation to "help" and "do good" with our research. However, impact is more specific than simply doing good. At a high level, impact in healthcare means that your research will ultimately make *a difference in healthcare practice*.

When we speak of impact in healthcare, we typically imply that your research does not only provide generalisable knowledge for the scientific constructs studied (e.g., communication, design processes, machine learning algorithms) but that it also changes artefacts or processes to improve people's lives, society, or the economy within the health domain. Thus, one of the obvious purposes of conducting studies in healthcare is to identify changes that can transform practice (e.g., Pronovost et al, 2010).

One of the most obvious reasons to consider impact at every level of a project is to secure funding. Often funding agencies expect you to explicitly specify the intended impact of your research during the proposal process. This is not simply a hand-waving exercise. Funding agencies are not only interested in knowing the implications of research, but also consider the impact of the research they fund. Some very good research ideas are ultimately not funded because there is no evidence of who would benefit from the research conducted and how that would be achieved.

Impact does not have to be localised to patients and healthcare services. One framework to guide your thinking about the importance and permutations of impact of research is the U.K.'s Engineering and Physical Sciences Research Council (EPSRC) impact framework (EPSRC, n.d.). This framework is meant to guide applicants and reviewers to consider the merits of more far-reaching effects of proposed research including: people, knowledge, society, and economy.

6.2   WHEN SHOULD I CONSIDER IMPACT?

In order to achieve these changes, impact should be considered early on in the research project process. Although this chapter comes last, impact should be the *first* thing you think about when embarking on a new research project. This does not mean you have to know exactly how that impact will be realised; but rather you should have the potential impact areas as an end goal in mind. You should know who and what will be affected by the findings you hope to elicit and present through your work.

As you begin your research project, it is good practice to sit down and write out these end-goals. The first set of end goals will relate to an increase in particular types of knowledge: this is the typical end-goal of research. It may be knowledge about behavioural theory or system design, but the implication is that you are advancing knowledge in your field. The second set of end goals will relate to the practice of healthcare. This is where you must take a step back from the minutiae of your particular research study and derive the benefits of your work for practice. These benefits could be realised in many different forms and levels: from a practice change or introduction of a new process or system within a hospital's department (see Chapter 5) to a change in national policy. An important step is not only to identify *who* will benefit from your research, but also how you will make that happen, i.e., you need to consider "pathways to impact." You must define the activities and then define strategies for working with other groups, such as clinical partners, to shorten the time between discovery and knowledge translation. We will discuss the varied ways you can have impact in the next section, but our point here is that it is important to consider this aspect from the beginning and regularly check whether or not you are fulfilling those end-goals when you make decisions regarding study design, implementation, and analysis.

6.3   WHAT TYPES OF IMPACT ARE FEASIBLE?

The impact of your research can take many different forms. Impact could be deep and narrow, or broad and shallow: you might impact a small group of patients in a deep way, or you may raise awareness of patient safety issues across a healthcare system.

In the following sections we have outlined six areas of potential impact: impact on researchers and practitioners; scientific knowledge; design of technology; practices; society; and economy. This is by no means an exhaustive list, but it describes some of the most common and successful forms of impact.

Figure 6.1: Dreaming of Impact. What different sorts of impact do you hope to make through your research? What evidence would demonstrate this impact?

6.3.1   IMPACTING RESEARCHERS AND PRACTITIONERS

An important aspect of conducting research is the impact it will have on the knowledge, skills, and experiences of researchers: these researchers may be the next generation of leaders in the field. This research pipeline needs to be maintained from school, to bachelor's degree, to Ph.D., to lecturer, to professor. Indications suggest that healthcare technology and patient safety issues are on the increase (e.g., the relatively recent implementation of Standard ISO/IEC 62366), which means this is a growth sector and more people with the right skills and experiences will be needed in the future.

Achieving a significant and lasting impact in healthcare practice is a daunting task, especially for a graduate student. Realising the impact the research is having on your development as a researcher and practitioner seems more manageable and is also important. Funders want to contribute to the development of highly skilled individuals who will eventually apply their skills to bring value to society and the economy. To put it crudely, human factors work in healthcare will be stunted if people with the right experiences and skill sets are not available to do the work.

One framework that can be used to articulate, assess, and manage the development of your skills and experience in research is the Vitae's Research Development Framework (Vitae, n.d.).

This is segmented into four main areas: (1) the knowledge, intellectual abilities, and techniques to do research; (2) the personal qualities and approach to be an effective researcher; (3) the knowledge of the standards, requirements, and professionalism to do research; and (4) the knowledge and skills to work with others and ensure the wider impact of research. All of these areas can be stretched and developed through the challenges of doing research in healthcare. These skills not only provide a good foundation for research, but they are also valuable and transferrable to other contexts. Examples include the high level of organisation needed to manage the governance procedures associated with healthcare research and the people skills needed to perform fieldwork in high-pressured environments.

Impact on researchers and practitioners, inside and outside healthcare, can be realised through education and training. For instance, Taneva and colleagues began to recognise that lasting change could really only happen through culture change—within the healthcare system, within the companies who create health technologies, and within the educational system that trains clinicians, health informaticians, and clinical engineers. They have been advocating for a culture of better design through various training and education programs. For example, Healthcare Human Factors runs a half-day course for clinicians on basic Human Factors principles (e.g., human limitations and biases, good design, and error prevention) in order to help them identify opportunities for improvement, and a separate course targeted toward designers and developers in healthcare.

Providing training opportunities to other researchers or students is possible through mechanisms such as courses or workshops at conferences such as the Conference on Human Factors in Computing Systems (CHI) or American Medical Informatics Association (AMIA). Disseminating your findings through these mechanisms ensures that other researchers and their goals benefit from your work.

6.3.2 IMPACTING SCIENTIFIC KNOWLEDGE

One of the main contributions of research in HCI and Computer-Supported Cooperative Work (CSCW) is the advancement of scientific knowledge regarding processes, artefact design, and policy implications. This is a natural outcome of fieldwork in healthcare. However, impacting scientific knowledge entails a great deal of effort, i.e., empirical results need to be translated into generalisable knowledge applicable to a number of health environments. It is good practice to point out the scope and limitations of your findings when presenting research. When it works well advances in this area can help improve the way we think about a processes and practices.

For instance, when undertaking a study of handovers in the U.K. (Volume 1: Chapter 5), one of the challenges that the researchers came across was that in some settings there were no obvious handover occurrences. This led the researchers to reflect on what they meant by the term *handover* (Wilson et al., 2009). Formal definitions of handover had suggested a single point of transition where responsibility is simultaneously relinquished by one party and accepted by another. An implication of this is that all information necessary for continuous safe care should be passed and received at this point in time. However, the researchers' observations suggested that it may be more beneficial to consider handover as a process, consisting of three components—transfer of responsibility, acceptance of responsibility, and the sharing of information about patients and tasks—and these components can occur in a variety of orders. In re-conceptualising handover, the researchers made a theoretical contribution with many different implications, including the design of IT systems to support handovers in a number of different contexts.

Likewise, in a different study of handover, Hilligoss and Moffatt-Bruce (2014) used their findings to make a theoretical contribution concerning communication in healthcare more generally, reflecting on the impact of paradigmatic communication (where information is organised into categories with context removed) and narrative communication (where information is organised into a story, with contextual information retained). Again, these conceptualisations of communication have implications for the design of IT systems.

Disseminating new scientific knowledge is commonly achieved through papers. Often the real contribution of a paper is not the specific empirical findings, but rather the implications of those findings and how they can be used in different contexts. A common challenge for graduate students is to explain how their findings expand our understanding of a phenomenon. They may confirm what we already know in a particular healthcare context, apply knowledge from another context to the healthcare domain, or, like the examples above, expand our understanding of a concept within healthcare.

This is another good reason why thinking about impact early on is necessary—you cannot have impact on knowledge without fully realising what we already know from prior work. You should do your due diligence and start with a literature review—oftentimes across multiple disciplines—in order to properly situate your own work and generalise and disseminate your findings to various interested parties.

6.3.3   IMPACTING THE DESIGN OF TECHNOLOGY

One obvious way to achieve impact is to deliver an artefact that transforms practice, based on the findings of fieldwork studies. For instance, we have already seen the MONARCA project, in Volume 1, that aimed to improve the quality of life for people with bipolar disease (Volume 1: Chapter 8), and the Spheres of Wellbeing project that aimed to improve the quality of life for female patients with borderline personality disorder (Volume 1: Chapter 9). Thinking more about international development, the PartoPen project seeks to make impact by trying to facilitate the application of WHO guidelines through technological support in Kenya (Volume 1: Chapter 12).

Whereas *it is possible* for a clinician or clinical team to adapt and develop a medical device for use within their own local practice (with only local IRB for quality improvement purposes but without going through any national regulatory approval process), it is required of researchers to engage with the national regulatory frameworks if wider use of the system is their goal. In the U.K., Mentis has had experience delivering impact at a local level by helping an in-house team develop technology for their hospital (O'Hara et al., 2014), but was restricted in testing the same technology in another hospital because it was not developed in-house there. These studies of innovative healthcare technologies required different approvals. However, Mentis has recently found it useful to partner with companies engaging in systems development. In these partnerships, Mentis provides the company with design ideas and the company provides systems that have been properly vetted by national authorities. In these cases, the HCI researcher can focus on conducting studies in the field as opposed to gaining device approval from regulatory agencies.

Anyone aiming to achieve impact through the development of innovative healthcare technologies needs to check whether or not the intended technology would be classed as a medical device. If it is considered a device, then you must determine what class of device it is and what the corresponding regulations are. If in doubt, talk with the regulators, and other specialists involved in relevant standards and regulatory requirements. In the U.S., your IRB, the Food and Drug Administration, as well as other regulatory agencies, are trained to make these determinations or at least can assist you in making these determinations. Chapter 5 in this volume discusses the design of intervention studies, outlines issues around medical device classification, and what needs to be considered in terms of ethical approval and governance procedures.

Within our own work, we have designed and deployed a number of technologies that have had impact on healthcare. As part of the study of handovers, Randell and her colleagues developed technology to support the work of a paediatric acute retrieval service (Wilson et al., 2010). A paediatric acute retrieval service is a mobile paediatric intensive care unit that assists general hospitals without specialist paediatric services and ensures safe and rapid transfer to a paediatric intensive care unit. The field studies allowed the researchers to understand the nature of the work of the retrieval service and their need for technology that was flexible, and which did not add to their work or distract their attention from the patient. Therefore, the researchers opted for a combination of digital pens/paper and shared displays to provide light-weight capture and sharing of information in (almost) real-time. As a consequence of the preliminary fieldwork, staff embraced the resulting system with enthusiasm and chose to buy the technology so as to continue using it after the study had ended.

Another example of a healthcare technology that has impacted practice is the Fall Prevention Tool Kit (FPTK) (Dykes et al., 2010). The FPTK, which tailors fall prevention interventions to address a patient's specific determinants of fall risk, was developed following a qualitative study to identify barriers and facilitators to fall risk communication and interventions in hospitals. When evaluated in a randomised trial, the FPTK was found to significantly reduce the rate of falls, preventing injury and pain for hospitalised patients.

Mentis and her collaborators from Microsoft Research Cambridge and Lancaster University conducted a significant number of observations in various operating rooms before they embarked on the design and development of a Kinect-based touchless imaging interaction system (Johnson et al., 2011; Mentis et al., 2012; Mentis and Taylor, 2013). Their fieldwork in interventional radiology, neurosurgery, and vascular surgery led to the successful design of a system subsequently deployed in the vascular surgery theatre of St. Thomas' hospital in London (O'Hara et al., 2014). During initial deployment, the research team conducted a series of fieldwork observations to better understand and explicate the impact that touchless interaction had on surgical practice. From this early work, their research findings and insights have been taken up by companies that have been deploying Kinect-based touchless systems to operating rooms around the world.

6.3.4   IMPACTING PRACTICE

Impact can also be achieved in healthcare by reviewing and assisting with practices associated with clinical care, such as device assessments, workplace assessments, quality improvement initiatives, and incident investigations. Thus, impact in healthcare does not always rely on the development of an artefact, but can simply be to inform work practices. In Section 6.4.2, we describe further how to go about translating and disseminating these ideas to healthcare administration. In this section we provide two examples of the form that impact on practice can take. The first example describes impact on device assessment for hospital procurement, which eventually led to a redesign of the products. The second example describes the assessment of an Emergency Department (ED) work-place environment to examine the impact of task interruptions on emergency physician workflow in order to change their work practices and floor layout.

Impacting Practice: Device Assessment

In 2004, University Health Network (UHN), one of the largest teaching hospitals in North America, had just embarked on a procurement project to replace all of its large volume intravenous (IV) infusion pumps. Infusion pumps are ubiquitous devices in a hospital, and this purchase not only represented a multi-million dollar investment, but also the potential to significantly impact the workflow of almost every one of UHN's thousands of nurses. This impacts the practice of IV administration in hospital.

Part of the hospital's procurement process typically is ensuring that each device meets vendor specifications and conforms to safety standards, but in this instance, the hospital took another step. In addition to their traditional process, the hospital-embedded human factors team (of which our co-authors Taneva and Chagpar are part) decided to engage in the procurement process and to conduct a comparative usability evaluation of contender products. This added rigour was done in part to acknowledge the financial and user adoption significance, but also to address the potential safety risks the new pumps could introduce. This evaluation was the first to make use of the hospital's new dedicated, state-of-the-art usability labs. This impacts the practice of procurement in hospital.

While the team was able to conduct the testing through a number of high risk and high frequency scenarios with nurses, the outcomes of the evaluation revealed less than optimal devices in terms of safety and usability. All three of the short-listed pumps had significant safety issues. UHN decided not to proceed with the procurement. Instead, as part of the analysis, the human factors team gave each manufacturer a detailed report of the findings, including video highlights of the critical errors that nurses repeatedly made with their products. Of the three manufacturers who received reports on the usability testing results, one of them saw value beyond the procurement evaluation. They were embarking on the development of a next-generation infusion pump and, based on the insight they received, the vendor requested the team's help with the new pump design. Over the next three years, the human factors team conducted 13 formal and ad-hoc usability tests to help the development of the new infusion pump. This work began broader collaborations between the team and the medical device industry. Although the team was active in patient safety initiatives internal to the hospital, they recognised that real systemic change could only happen if they worked directly with manufacturers to create better health technologies. That way, instead of mitigation strategies and work-arounds to address poor product design, they could influence the design of next generation products directly. This broadly impacts the practice of IV administration.

Impacting Practice: Workplace Culture and Environment

Researchers from MedStar Health's National Center for Human Factors in Healthcare have examined the impact of interruptions in the ED. Task interruptions have long been recognised as a problem in healthcare and are particularly challenging in the ED where interruptions may be necessary for the effective delivery of care (Berg et al., 2013; Chisholm et al., 2000; Westbrook et al., 2010). In an effort to assess the nature of interruptions in the ED and to improve healthcare processes the research team, led by Ratwani, conducted a field study of the ED focused on understanding the characteristics of interruptions and recommending improvements to reduce their disruptiveness.

The team developed a mobile application to track when physicians were interrupted, who they were interrupted by, and where the interruptions occurred in the ED (Fong et al., 2014). The data showed that physicians were most commonly interrupted at their central workstations while on computer-based tasks and that they were commonly interrupted by nurses and technicians. Importantly, while the length of each single interruption was relatively short, the interruptions tended to occur in clusters where, once the physician was interrupted, several other interruptions followed.

The field study and resulting data were leveraged to provide the following recommendations for ED physician practice:

- Tasks that are prone to error due to interruption should not be performed at the common workstation where interruptions are frequent. These tasks, such as computerised medication ordering, should be completed in more protected spaces like the patient room.

- To provide flexibility to physicians in where they place their computer orders, laptops, workstations on wheels, or other mobile technology should be employed.

- Clinical staff should work to develop cognitive strategies to mitigate the disruptiveness of interruptions. This may include memory strategies to remember the particular task being worked on as well as using external cues to track task progress.

The field study and resulting recommendations have had an impact on physician performance and workflow, as well as on patient safety and care quality.

6.3.5   IMPACTING SOCIETY

Impacting society is ambitious for any single research project, but even single research projects can play their part in broader movements that aim to impact policies, practices, and culture at a national and international level. Impacting society can also take the form of public engagement where researchers aim to raise debate and awareness about important issues (e.g., the role of human factors in reducing avoidable patient harm) and to educate and inspire the next generation about exciting topics (e.g., HCI and human factors).

One way for researchers to achieve impact on society is through active participation in policy, government, and hospital committees. For example, members from the Healthcare Human Factors team at UHN sit on a number of hospital patient safety committees, as well as standards committees of the Association for the Advancement of Medical Instrumentation and the Canadian Health Informatics Association, helping shape the level of rigour required for medical technology development and evaluation. In the U.K., the Clinical Human Factors Group (www.chfg. org), made up of healthcare professionals, managers, researchers, and human factors experts, have succeeded in persuading the Department of Health to establish a working group to identify routes to embed human factors thinking in the National Health Service and have provided written and verbal evidence to the Parliamentary Select Committee on Patient Safety. Similarly, in the U.S., researchers from the National Center for Human Factors in Healthcare have provided advice on electronic health record usability policy.

Researchers have also achieved impact on society at national and international levels by doing research that is broadly disseminated through academic and non-academic channels and captures media attention. A great example is Atul Gawande and his book, *The Checklist Manifesto: How to Get Things Right*, which has helped direct and raise the profile of an international movement to promote the use of checklists in surgery. In his studies of eight hospitals Gawande found that major postsurgical complications dropped by 36 percent in the six months after the checklist was introduced, and deaths fell by 47 percent (Henig, 2009). Following this work, the World Healthcare Organisation (WHO) launched a Surgical Safety Checklist for use in operating room environments. This impact on work practices and checklists is well within the realms of human factors, although a surgeon has led it. This work transcended the theatre and normal research reportage to impact the popular media. This case study invites us to reflect on what has led to such widespread attention and impact, particularly when smaller studies might feel they struggle to be heard. Some answers are offered by Greenhalgh et al. (2004) who report on the factors that impact the diffusion of innovation in service organisations. One of the clear messages from this paper is that the successful and widespread diffusion of innovation involves many more factors than just a good idea.

Impact on society is likely to be more effective if practitioners, patients, and the public not only *participate* in the research, but are *involved* and *engaged* with it, too (these terms are defined in Chapter 1). An informed and educated society is arguably valuable in its own right. To achieve impact, in terms of raising awareness and informing people about HCI and human factors issues in healthcare, you can run public engagement activities, blogs, make videos, and hold seminars to inform and inspire people about the research topics and studies we are interested in. For example, *Microwave Racing* is a YouTube video that won a CHI Video Showcase Award. This video has been shown in schools and has been used as an activity at science festivals to inspire children and adults to think about interaction design (Furniss, 2010; Black et al., 2012). Along a similar vein, TED talks on relevant topics have been widely shared among the general public, e.g., Brian Goldman's talk, *Doctors Make Mistakes. Can we talk about that?* (Goldman, 2012). Thought provoking videos have also been posted to YouTube, e.g., Martin Bromiley's *Just a Routine Operation* (Laerdal Medical, 2011). Of course society may be sliced into smaller, more manageable segments, so public engagement messages that are pertinent to different groups can be more focused and targeted. For example, raising awareness of tragic cases of the *second victim* amongst healthcare professionals, like Kimberly Hiatt who took her own life after a dosing error that killed a baby, might encourage more support for staff after an incident (Dekker, 2013). Again, when we think about successful diffusion and impact it is not only about what the message is but also about who the person delivering that message is and who the audience is, how the message is delivered in terms of the style and channel, when it is delivered in terms of whether an audience will be receptive or not, and why the audience should listen and care about the message.

> **Tip:** When planning to engage with different audiences, it is advisable to try to piggyback on existing events, social forums, professional bodies, mailing lists, science festivals, charities, patient support groups, seminar series, and publication outlets as they will already have a community and a mechanism to engage with them.

6.3.6   IMPACTING THE ECONOMY

There are varied ways of impacting the economy and economic arguments for improving aspects of healthcare can be persuasive. For example, promoting a healthy workforce, saving litigation costs, reducing hospital bed stays, developing medical innovations and marketable devices are just some examples of impacting economy. This could be achieved, for example, by developing intellectual property, and spinout companies to bring new equipment and services to market. These are achievable but could appear daunting to non-entrepreneur types.

We have already seen how human factors can impact practice above (e.g., through procurement, device design, and reducing interruptions), which can have direct and indirect effects on costs. Another way to impact economy is to reduce the need for inpatient care and clinical visits. One area where this can be achieved is the surgical domain, through increasing the use of laparoscopic (keyhole) surgery. Mentis has been conducting fieldwork in operating rooms in order to support laparoscopic procedures through the design of enhanced endoscopic video displays as well as the design of new surgical training simulators (Mentis et al., 2014). In addition, Mentis has recently embarked on fieldwork on the assessment of movement ability in patients with Parkinson's disease in order to identify technology needs for home-based monitoring and medication decision making. This work supports impact in economy by reducing the time and money associated with frequent clinical visits by those with chronic and progressive diseases.

6.4   HOW CAN WE TRANSFER THE FINDINGS OF FIELDWORK TO ACHIEVE IMPACT?

Fieldwork in healthcare can certainly have impact in a number of different ways, but in order to be successful in translating your results to improve practice, you need to consider how you present your work. This involves, first, identifying who needs to be made aware of your findings or knowledge. This often includes professional societies of your study population (e.g., surgical societies to present research on new surgical training simulations) or groups associated with policy and administrative decision making. The key here is that, in order to *transfer* your findings and disseminate your recommendations, you must often *translate* them. This involves tapping into key audiences in terms of language, culture, and values. For example, there is a big difference in the way you would present usability findings to HCI and human factors researchers, as opposed to clinicians who use these systems, manufacturers that build the systems, policy makers that shape the use of systems, regulators who approve the systems, and insurance companies that pay for the use of systems. However, broad dissemination of results is a key factor in having impact. If you only present work at an HCI conference, you are less likely to get the attention of people that could benefit from it, i.e., its value is not realised. This is where partners and collaborators in the field or in the industry can improve the likelihood of impact.

6.4.1   WRITING FOR DIFFERENT AUDIENCES

Different communities have different cultures with informal rules and value systems. This is easier to manage when you have someone alongside you well versed in that culture. For instance, writing results for a medical journal is very different from writing for a human factors journal. For one, the average length of most medical journal articles is five pages. It is difficult to present the findings from a qualitative field study in detail in such a small space. Moreover, every discipline has evolved its own values and expectations related to what constitutes a valid research method, what literature to cite and how, how to construct a credible argument, or how to structure an abstract. To be successful, it is essential to be aware of, and respond to, different expectations. This is where having a healthcare professional as a co-author will give you insights on what data to include and how to best present methods and take-away points.

When writing for healthcare, it is important to consider who is, in fact, the most relevant audience. Depending on the nature of a study, it may be more appropriate to report in a generalist healthcare journal or one focusing on a particular clinical speciality. It is important to be aware of the detailed focus of particular journals (e.g., medicine or healthcare) and their level of interest in fieldwork studies. Clinicians' interest in evidence-based medicine, and a focus on outcomes-related research, can make reporting to clinical audiences challenging. The idea that process and outcome may not be closely related, when evidence-based medicine relies on understanding those differences, may be difficult to impart, and a failure to report statistics makes some clinicians consider HCI research less credible than quantitative, outcomes-based research (Concato et al., 2000). Among clinicians, positivism is still a dominant philosophy. It is important to justify the value of the questions being addressed and the legitimacy of the approach being taken. One useful strategy is to contact the editor of a journal you consider a good fit and ask whether your work fits the scope of the journal. In addition, look for other qualitative articles in that journal to see how they are structured and how the findings are presented. As qualitative research becomes better established in healthcare, there will be a larger body of literature to invoke and to model new research on.

However, if you wish to impact practice, you should not restrict your dissemination just to academic health and medical journals. There are a range of professional journals for the different groups that make up the healthcare workforce. For example, in the U.K., professional journals include the *Nursing Times*, the *Bulletin of the Royal College of Surgeons*, and the *Health Service Journal*, which is aimed at healthcare managers. Again, these journals have their own particular style which you should familiarise yourself with; the articles are much shorter than those in academic journals, but they also have a much larger readership than academic journals.

We also need to be aware of professional sensitivities when presenting findings of qualitative studies to an audience whose training has emphasised perfect performance (just think of the language of "never events": if it should never happen, how can we talk about such events and their causes?). The idea that the socio-technical context can influence performance, and that performance is not perfect, can be challenging for many professionals. The fact that this perspective is alien to many clinicians makes it all the more important to communicate findings in the healthcare context, so as to inform future practice.

In terms of the pragmatics of writing or presenting for different audiences, it is important to also be aware of different practices; for example,

- some HCI journals have no restrictions on paper length, so that papers commonly exceed 10,000 words, whereas most health journals have fixed word limits, often 4,000 or fewer;

- HCI journals typically accept only one kind of paper (original research) some health journals accept many kinds (research, case studies, viewpoints, etc.); and

- many HCI conferences produce archival proceedings based on the submission of full papers, while health conferences are more likely to be based on submission of abstracts, or feature invited speakers.

These other forms of submissions present a unique opportunity for disseminating results to health and medical journals, for instance by presenting review articles or opinion pieces on new technological directions. Thus, work that might not be seen as a contribution to the HCI field may be seen as unique and novel to health and medical fields. The starting point is not to assume that any two communities have shared practices in terms of disseminating the findings of research, and to become familiar with the variety of practices.

Increasingly, healthcare researchers are using social media such as Twitter and blogs to not only connect with other researchers but also to connect with a broader audience of healthcare professionals and organisations, providing another way to communicate your research. If your project gets media attention then add this to the case to demonstrate impact. If seminars and public engagement events are held, then survey who is there, get feedback, and evaluate whether they thought it was interesting and what they learned. Keep emails and social media messages that provide you with feedback or that request more information or involvement, as this will help provide evidence for impact.

> **Tip:** When developing a case to demonstrate impact, direct quotations from patients and members of the public can be a powerful and persuasive form of impact, particularly when combined with more quantitative approaches, which together show broader and deeper impact.

6.4.2 SPEAKING WITH ADMINISTRATIVE DECISION MAKERS

Health services worldwide are complex organisations: simultaneously hierarchical and distributed. Under constant pressure, there is a tendency to focus on survival and dealing with the immediate problems rather than being able to step back and take a strategic view on change. In addition, healthcare is often heavily regulated and implementing new practices or procedures can be met with resistance due to uncertainty of the regulatory environment.

Findings that you think are indicative of why your system should be used or the reason a policy should be changed may not resonate with the type of audience that is really in the position to make such decisions. This is often the case when the decision making power lies in the hands of healthcare administration. Although you may show through your research that an increase in nurse communication during shift changes through your technological intervention has led to better patient care and reduced risk of medical error, you may find that a better route would be to present the cost savings in terms of worker time and number of patients that can be seen as a *result* of the improved communication. Healthcare administrators, in particular, are generally concerned with the cost of care and the efficiencies of delivering care. If you do not measure outcome variables such as time and patient turnover, you are likely going to find it difficult to make the argument to the administration that your technology is worth the time and money it would take to implement it on a grander scale. However, it is also important to stay alert to what is happening in the healthcare environment, in terms of the political and policy drivers for change. At the time of writing, hospitals in the U.K. are responding to the recommendations of the Francis Inquiry into major failings in patient care at one hospital Trust (The Mid Staffordshire NHS Foundation Trust Public Inquiry, 2013). This case raises the profile of quality and safety of patient care across the entire NHS and could mean that technologies with the potential to positively impact patient care receive more attention than those that simply reduce costs.

Getting healthcare administrators on board can be particularly difficult if the people proposing change are perceived as outsiders. Yet, the distribution of responsibilities across staff and teams, and the transient nature of many roles (e.g., trainee doctors, agency nurses) makes it difficult to identify and work with individuals who have the power and motivation to effect change. This emphasises the importance of engaging with healthcare managers early on, to explain to them the purpose of your study and its potential impact, and to ideally find stakeholders who genuinely care about the issues under study. By engaging with them early on, you will have a better understanding of their concerns and will be able to think about how your research and subsequent recommendations could address these concerns. It may be that the organisation has introduced technology that they would be keen for you to evaluate. Alternatively, they may be doing their own evaluation, in which case it is up to you to explain to them the added value that a field study could provide. For example, as part of the study of handover described previously, one of us was invited to contribute to the evaluation of technology that a hospital was introducing to support handover. Although invited, the study presented the same challenges as in any other setting, but we had the opportunity to contribute to discussions amongst the project management team, who consisted mainly of hospital managers and IT staff, with some representation of clinical staff. The challenges for the managers and IT staff were getting a realistic understanding of how the work took place and how the technology was used, and they valued the fact that we could provide this kind of input.

6.5　SUMMARY

Impact can be achieved in various ways, such as: working on problems that have already been identified by healthcare staff, so that they have a sense of "ownership" of the problem; working with clinical champions who have the power to effect change within the organisation; and working with umbrella organisations (such as professional bodies, patient groups, or health-related charities) who can disseminate findings and interventions widely. All of these strategies suggest the need to build connections and frame your work with an eye toward impact early on.

Finally, it may be necessary to accept that, while you may have achieved impact in terms of advancing knowledge and may be making a theoretical contribution, impacting healthcare practice is a challenging process that may not be achievable within the timeframe of the typical project. This is not due to the nature of the study; achieving impact is a slow process and is typically achieved through undertaking a body of research, not just one study, and so this may not be a realistic aim for a Ph.D. student. Often real, long-lasting impact occurs over the course of a program of work and years of relationship building. That does mean, though, that every project you engage in is a stepping-stone toward real impact in healthcare.

6.6　CONCLUSION

Throughout this book we have encountered important topics and issues to consider for conducting fieldwork in healthcare. We began with looking at ethics, governance, and PPI before looking at readying the researcher, which involved reflecting on your role and identity, the emotional challenges that might arise, and remaining safe. We then discussed establishing and maintaining relationships in healthcare that are essential for conducting your studies. This led into looking at the practicalities of data collection, and then on to the practicalities of conducting intervention studies that aim to develop and deploy novel technologies in context. This sixth and final chapter has expanded the notion of impact and given examples of how we, as researchers doing HCI, human factors, and related research, can have an impact on healthcare.

These six chapters have built on the experience of many different researchers. Some of these are documented in the case studies of Volume 1. These give rich reflections on the successes and challenges encountered across a variety of studies. Indeed, one of the attractions of working in healthcare is the rich depth and variety of phenomena, which fuels continual opportunities for learning and potential for impact. You will experience new pockets of healthcare you were not aware of, new research challenges, new ethical issues, new communities to reach out to and engage with, and new technology that needs development, direction, or evaluation. There is such a variety of contexts—many different diseases, disabilities, treatments, procedures, personalities, patients, and professionals; new technological developments, and a continuous drive to reduce costs and improve standards. Our professions need to play an increasingly important role in patient safety and care quality. After all, people's lives are impacted by the design, technologies and services on a daily basis. Healthcare is intimately connected to life.

There is a lot to learn, and continue to learn. These companion volumes will provide a welcome complement to your journey, whether you are just starting out or are some way down the road.

References

AHIMA. (March 2011). Security Audits of Electronic Health Information (Updated). *Journal of AHIMA* 82(3), 46–50.

Als, A. B. (1997). The Desk-Top Computer as a Magic Box: Patterns of Behavior Cconnected with the Desk-Top Computer; GPs' and patients' perceptions. *Family Practice* 14(1), 17–23. DOI: 10.1093/fampra/14.1.17.

Arthur, S. and Nazroo J. (2012). *Designing Fieldwork Strategies and Materials. Qualitative research practice: A guide for social science students and researchers.* J. Ritchie and J. Lewis. London, SAGE Publications: 109–137.

Attfield, S. J., Adams, A., and Blandford, A. (2006). Patient Information Needs: Pre-and Post-Consultation. *Health Informatics Journal*, 12(2), 165–177. DOI: 10.1177/1460458206063811.

Bardram, J.E., Hansen, T.R., and Soegaard, M. (2006). AwareMedia: A Shared Interactive Display Supporting Social, Temporal and Spatial Awareness in Surgery. In *Proceedings from The Conference on Computer-Supported Cooperative Work.* New York, NY: ACM Press, 109–118.

Barry, C. A., Britten, N., Barber, N., Bradley, C., and Stevenson, F. (1999). Using Reflexivity to Optimize Tteamwork in Qualitative Research. *Qualitative health research*, 9(1), 26–44. DOI: 10.1177/104973299129121677.

Berg, B. L. (2004). *Qualitative Research Methods*, Fifth Edition., Allyn and Bacon.

Berg, L. M., Källberg, A.-S., Göransson, K. E., Ostergren, J., Florin, J., and Ehrenberg, A. (2013). Interruptions in Emergency Department Work: An Observational and Interview Study. *BMJ Quality and Safety*, 22(8), 656–663. doi:10.1136/bmjqs-2013–001967. DOI: 10.1136/bmjqs-2013-001967.

Bernard, H. R. (2002). *Research Methods in Anthropology.* Walnut Creek, CA, Altamira press.

Bhutta, Z. A. (2002). Ethics in International Health Research: A Perspective from the Developing World. *Bulletin of the World Health Organisation*, 80, 114–120.

Black, J., Furniss, D., Myketiak, C., Curzon, P., and McOwan, P. (2012). Microwave Racing: An Interactive Activity to Enthuse Students About HCI. *The Contextualised Curriculum Workshop at CHI 2012.*

Borbasi, S., Jackson, D., and Wilkes, L. (2005). Fieldwork in Nursing Research: Positionality, Practicalities and Predicaments. *Journal of advanced nursing*, 51(5), 493–501. DOI: 10.1111/j.1365-2648.2005.03523.x.

Booth, W. (1998). Doing Research with Lonely People. *British Journal of Learning Disabilities*, 26(4), 132–134. DOI: 10.1111/j.1468-3156.1998.tb00068.x.

Brain, J., Schofield, J., Gerrish, K., Mawson, S., Mabbott, I., Patel, D., and Gerriish, P. (2011). *A Guide for Clinical Audit, Research and Service Review*. Healthcare Quality Improvement Partnership.

Britten, N. (2006). *Qualitative interviews. Qualitative Research in Health Care*, Third Edition, C. Pope and N. Mays. Malden, Mass., Blackwell Publishing: 12–20.

Campbell M, F. R. (2000). Framework for Design and Evaluation of Complex Interventions to Improve Health. *BMJ*, 321 (7262), 694–696. DOI: 10.1136/bmj.321.7262.694.

Campbell, M., Fitzpatrick, R., Haines, A., Kinmonth, A. L., Sandercock, P., Spiegelhalter, D., and Tyrer, P. (2000). Framework for design and evaluation of complex interventions to improve health. *BMJ: British Medical Journal*, 321(7262), 694.

Catchpole, K. R., Giddings, A. E., Wilkinson, M., Hirst, G., Dale, T., and de Leval, M. R. (2007). Improving Patient Safety by Identifying Latent Failures in Successful Operations. *Surgery*, 142(1), 102–110. DOI: 10.1016/j.surg.2007.01.033.

CDC (n.d.). *U.S. Public Health Service Syphilis Study at Tuskegee* http://www.cdc.gov/tuskegee/timeline.htm. Accessed 18th February 2014.

Chen, E. S. and Cimino, J. J. (2003). Automated Discovery of Patient-Specific Clinician Information Needs Using Clinical Information System Log Files. *Proceedings of the American Medical Informatics Association 2003 Annual Symposium* (AMIA 2003), 145–149.

Chisholm, C. D., Collison, E. K., Nelson, D. R., and Cordell, W. H. (2000). Emergency Department Workplace Interruptions: Are Emergency Physicians "Interrupt-Driven" and "MultiTasking"? *Academic Emergency Medicine*, 7(11), 1239–1243. DOI: 10.1111/j.1553-2712.2000.tb00469.x.

Concato, J., Shah, N., and Horwitz, R. I. (2000). Randomized, Controlled Trials, Observational Studies, and the Hierarchy of Research Designs. *New England Journal of Medicine*, 342(25), 1887–1892. DOI: 10.1056/NEJM200006223422507.

Cutcliffe, J. R., and Ramcharan, P. (2002). Leveling the Playing Field? Exploring the Merits of the Ethics-as-Process Approach for Judging Qualitative Research Proposals. *Qualitative Health Research*, 12(7), 1000–1010. DOI: 10.1177/104973202129120313.

Davis, H. (2001). The Management of Self: Practical and Emotional Implications of Ethnographic Work in a Public Hospital Setting. In Gilbert, K. (Ed.). *The Emotional Nature of Qualitative Research*, 37–61. CRC Press.

Dekker, S. (2013). *Second Victim: Error, Guilt, Trauma, and Resilience.* CRC Press. DOI: 10.1201/b14797.

Dell, N., et al. 2012. "Yours is Better!" *Participant Response Bias in HCI.* CHI 2012. DOI: 10.1145/2207676.2208589.

de Melo, L. P., Stofel, N. S., Gualda, D. M. R., and de Campos, E. A. (2014). Nurses' Experiences of Ethnographic Fieldwork. *Nurse researcher*, 22(1), 14–19. DOI: 10.7748/nr.22.1.14.e1243.

Deparment of Health. (2005). *Research Governance Framework for Health and Social Care* (2nd Edition ed.). London: Deparment of Health.

Dykes, P., Carroll, D., Hurley, A., Lipsitz, S., Benoit, A., Chang, F., Meltzer, S., Tsurikova, R., Zuyov, L., and Middleton, B. (2010). Fall Prevention in Acute Care Hospitals: A Randomized Trial. *JAMA*, 304(17), 1912–1918. DOI: 10.1001/jama.2010.1567.

Emanuel, E. J., Wendler, D., and Grady, C. (2000). What Makes Clinical Research Ethical? *JAMA*, 283(20), 2701–2711. DOI: 10.1001/jama.283.20.2701.

Emanuel, E. J., Wendler, D., Killen, J., and Grady, C. (2004). What Makes Clinical Research in Developing Countries Ethical? The Benchmark of Ethical Research. *The Journal of infectious diseases*, 189 (5), 930–7. DOI: 10.1086/381709.

Emerson, R. M., Fretz, R. I., and Shaw, L. L. (1995). *Writing Ethnographic Fieldnotes*. University of Chicago Press. DOI: 10.7208/chicago/9780226206851.001.0001.

EPSRC (no date). Impact—Guidance for Applicants and Reviewers. http://www.epsrc.ac.uk/funding/howtoapply/preparing/economicimpact/. Accessed July 11, 2014.

Faulkner et al. (2013). Exploring the Impact of Public Involvement on the Quality of Research: Examples. *NIHR INVOLVE,* July 2013.

Finlay, L. (2002). "Outing" the Researcher: The Provenance, Process, and Practice of Reflexivity. *Qualitative health research*, 12(4), 531–545. DOI: 10.1177/104973202129120052.

Finlay, L. (2003). The Reflexive Journey: Mapping Multiple Routes. In L. Finlay ANDB. Gough (Eds.) *Reflexivity: A Practical Guide for Researchers in Health and Social Sciences*. Blackwell Publishing. DOI: 10.1002/9780470776094.

Finlay, L., and Gough, B. (Eds.). (2003). *Reflexivity: A Practical Guide for Researchers in Health and Social Sciences*. Blackwell Publishing. DOI: 10.1002/9780470776094.

Fisher, A. A., and Foreit, J. R. (2002). *Designing HIV/AIDS Intervention Studies: An Operations Research Handbook*. New York: The Population Council Inc.

Fitzgerald, M. C., Gocentas, R., Dziukas, L., Cameron, P. A., Mackenzie, C. F., and Farrow, N. C. (2006). Using Video Audit to Improve Trauma Resuscitation—Time for a New Approach. *Canadian Journal of Surgery* 49(3), 208–211.

Flanagan, J. C. (1954). The Critical Incident Technique. *Psychological bulletin*, 51(4), 327. DOI: 10.1037/h0061470.

Fong, A., Meadors, M., Batta, N., Nitzberg, M., Hettiner, A., and Ratwani, R. (2014). Identifying Interruption Clusters in the Emergency Department. *Proceedings of Human Factors and Ergonomics Society 2014 Annual Meeting*. DOI: 10.1177/1541931214581135.

Frost, M., Doryab, A., and Bardram, J. E. (2013). Supporting Disease Insight through Data Analysis: Refinements of the MONARCA Self-Assessment System. *Ubicomp '13 The 2013 ACM Conference on Ubiquitous Computing*. New York, NY, USA: ACM. DOI: 10.1145/2493432.2493507.

Furniss, D. (2010). *Microwave Racing*. http://www.youtube.com/watch?v=Bzy5hVvbei8 Uploaded November 30, 2010. Accessed July 11, 2014.

Gilbert, K. (Ed.). (2001). *The Emotional Nature of Qualitative Research*. CRC Press.

Gilbert, K. R. (2001a). Introduction: Why Are We Interested in Emotions? In Gilbert, K. (Ed.). *The Emotional Nature of Qualitative Research*, 3–15. CRC Press.

Gilbert, K. R. (2001b). Collateral Damage? Indirect Exposure of Staff Members to the Emotions of Qualitative Research. In Gilbert, K. (Ed.). *The Emotional Nature of Qualitative Research*, 147–161. CRC Press.

Glaser, B. G. and Strauss, A. L. (1967). *The Discovery of Grounded Theory: Strategies for Qualitative Research*. New York, Aldine Publishing Company.

Glushko, A. (2013). Participatory Design in Healthcare: Patients and Doctors Can Bridge Critical Information Gaps, *UX Magazine*, article no. 1028. Retrieved from http://uxmag.com/articles/participatory-design-in-healthcare.

Gough, B. (2003). Deconstructing Reflexivity. In L. Finlay and B. Gough (Eds.) *Reflexivity: A Practical Guide for Researchers in Health and Social Sciences*. Blackwell Publishing. DOI: 10.1002/9780470776094.ch2.

Greatbatch, D., Heath, C., Campion, P., and Luff, P. (1995). How Do Desk-Top Computers Affect the Doctor-Patient Interaction. *Family Practice* 12(1), 32–36. DOI: 10.1093/fampra/12.1.32.

Greenhalgh, T., Robert, G., Macfarlane, F., Bate, P., and Kyriakidou, O. (2004). Diffusion of Innovations in Service Organisations: Systematic Review and Recommendations. *Milbank Quarterly*, 82(4), 581–629. DOI: 10.1111/j.0887-378X.2004.00325.x.

Goldman, B. (2012). Doctors Make Mistakes. Can We Talk about That? https://www.youtube.com/watch?v=iUbfRzxNy20 Uploaded: January 25, 2012. Accessed: July 11, 2014.

Gurses, A. P., Xiao, Y., and Hu, P. (2009). User-Designed Information Tools to Support Communication and Care Coordination in a Trauma Hospital. *Journal of Biomedical Informatics* 42(4), 667–677. DOI: 10.1016/j.jbi.2009.03.007.

Hammersley, M. and Atkinson, P. (1995). *Ethnography: Principles in Practice*. London: Routledge.

Hansen, T. R., Bardram, J. E., and Soegaard, M. (2006). Moving Out of the Lab: Deploying Pervasive Technologies in a Hospital. *Pervasive Computing*, 5 (3), 24–31. DOI: 10.1109/MPRV.2006.53.

Health Research Authority. (2013). Defining Research: NRES Guidance to Help You Decide if Your Project Requires Review by a Research Ethics Committee. Ref: 0987 December 2009 (rev. April 2013) http://www.hra.nhs.uk/documents/2013/09/defining-research.pdf.

Henig, R.M. (2009). *A Hospital How-To Guide That Mother Would Love*. http://www.nytimes.com/2009/12/24/books/24book.html?_r=0. Published December 23, 2009. Accessed July 11, 2014.

HHS (n.d.) *The Nuremberg Code*. http://www.hhs.gov/ohrp/archive/nurcode.html. Accessed 18th Feb 2014.

Hilligoss, B. (2014). Fieldwork and Challenges of Access. In D. Furniss, et al. (Eds.), *Fieldwork for Healthcare: Case Studies Investigating Human Factors in Computing Systems*, vol. 1, Morgan & Claypool Publishers. DOI: 10.2200/S00552ED1V01Y201311ARH005.

Hilligoss, B. and Moffatt-Bruce, S. (2014). The Limits of Checklists: Handoff and Narrative Thinking. *BMJ Qual Saf.* Published Online First: 2 April 2014. DOI:10.1136/bmjqs-2013-002705.

Hindmarsh, J. and Pilnick, A. (2002). The Tacit Order of Teamwork: Collaboration and Embodied Conduct in Anesthesia. *Sociological Quarterly* 43(2), 139–164. DOI: 10.1111/j.1533-8525.2002.tb00044.x.

Hindmarsh, J. and Pilnick, A. (2007). Knowing Bodies at Work: Embodiment and Ephemeral Teamwork in Anaesthesia. *Organisation Studies* 28(9), 1395–1416. DOI: 10.1177/0170840607068258.

Holloway, I. and Wheeler, S. (2013). *Qualitative Research in Nursing and Healthcare*. John Wiley & Sons.

Institute for Clinical Evaluative Sciences. (1999). Focus Groups in Health Services Research at ICES. *ICES Publication* No. 99-02-TR. Ontario, Canada.

Jalloh, O. B. and Waitman, L. R. (2006). Improving Computerized Provider Order Entry (CPOE) Usability by Data Mining Users' Queries from Access Logs. *Proceedings of the American Medical Informatics Association 2006 Annual Symposium* (AMIA 2006), 379–83.

Johnson, R., O'Hara, K., Sellen, A., Cousins, C., and Criminisi, A. (2011). Exploring the Potential for Touchless Interaction in Image-Guided Interventional Radiology. In *Proceedings of the SIGCHI Conference on Human Factors in Computing Systems*, New York, NY: ACM Press, 3323–3332. DOI: 10.1145/1978942.1979436.

Kane, B. and Luz, S. (2006). Multidisciplinary Medical Team Meetings: An Analysis of Collaborative Working with Special Attention to Timing and Teleconferencing. *Computer Supported Cooperative Work* (CSCW) 15, 501–535. DOI: 10.1007/s10606-006-9035-y.

Kane, B. T., Toussaint, P. J., and Luz, S. (2013). Shared Decision Making Needs a Communication Record. *Proceedings of the ACM 2013 Conference on Computer Supported Cooperative Work and Social Computing* (CSCW 2013), 79–90.

Kaplan, B. (2001). Evaluating informatics applications—some alternative approaches: theory, social interactionism, and call for methodological pluralism. *International Journal of Medical Informatics*, 64(1), 39-56. DOI: http://dx.doi.org/10.1016/S1386-5056(01)00184-8.

Karnieli-Miller, O., Strier, R., and Pessach, L. (2009). Power Relations in Qualitative Research. *Qualitative Health Research*, 19(2), 279–289. DOI: 10.1177/1049732308329306.

Khanlou, N. and Peter, E. (2005). Participatory Action Research: Considerations for Ethical Review. *Social Science and Medicine*, 60 (10), 2333–2340. DOI: 10.1016/j.socscimed.2004.10.004.

Kitzinger, J. (2006). *Focus groups. Qualitative Research in Health Care*, Third Edition, C. Pope and N. Mays. Malden, Mass., Blackwell Publishing: 21–31.

Klasnja, P. and Pratt, W. (2011). Healthcare in the Pocket: Mapping the Space of Mobile-Phone Health Interventions. *Journal of Biomedical Informatics*. Aug. DOI: 10.1016/j.jbi.2011.08.017.

Klasnja, P., and Pratt, W. (2012). Healthcare in the pocket: Mapping the space of mobile-phone health interventions. *Journal of Biomedical Informatics*, 45(1), 184-198. DOI: 10.1016/j.jbi.2011.08.017.

Koh, H. K. and Sebelius, K. G. (2010). Promoting Prevention through the Affordable Care Act. *New England Journal of Medicine*, 363(14), 1296–1299. DOI:10.1056/NEJMp1008560.

Kusunoki, D. S., Sarcevic, A., Weibel, N., Marsic, I., Zhang, Z., Tuveson, G., and Burd, R. S. (2014). Balancing Design Tensions: Iterative Display Design to Support ad hoc and Interdisciplinary Medical Teamwork. *Proceedings of the ACM SIGCHI 2014 Conference on Human Factors in Computing Systems* (CHI 2014), 3777–3786. DOI: 10.1145/2556288.2557301.

Kusunoki, D. S., Sarcevic, A., Zhang, Z., and Burd, R. S. (2013). Understanding Visual Zttention of Teams in Dynamic Medical Settings through Vital Signs Monitor Use. *Proceedings of the ACM 2013 Conference on Computer Supported Cooperative Work and Social Computing* (CSCW 2013), 527–540. DOI: 10.1145/2441776.2441836.

Laerdal Medical (2011). Just A Routine Operation. https://www.youtube.com/watch?v=Jzlvgt-PIof4. Uploaded: July 6, 2011. Accessed: July 11, 2014.

LeCompte, M. D. and Schensul, J. J. (Eds.). (1999). *Analyzing and Interpreting Ethnographic Data.* Rowman Altamira.

Lewis, J. (2012). *Design Issues. Qualitative Research Practice: A Guide for Social Science Students and Researchers.* J. Ritchie and J. Lewis. London, SAGE Publications: 47–76.

Lofland, J., Snow, D., Anderson, L., and Lofland, L. (2006). *Analyzing Social Settings.* Belmont, CA: Wadsworth Publishing Company.

Mackay, W. E. (1995). Ethics, Lies and Videotape… *CHI '95 Proceedings of the SIGCHI Conference on Human Factors in Computing Systems* (138–145). New York: ACM Press. DOI: 10.1145/223904.223922.

Mackenzie, C. F., Hu, P. F., Horst, R. L., and LOTAS Group. (1995). An Audio-Video system for Automated Data Acquisition in the Clinical Environment. *Journal of Clinical Monitoring and Computing* 11(5), 335–341. DOI: 10.1007/BF01616993.

Mackenzie, C. F. and Xiao, Y. (2003). Video Techniques and Data Compared with Observation in Emergency Trauma Care. *Quality and Safety in Health Care* 12, 51. DOI: 10.1136/qhc.12.suppl_2.ii51.

Marshall, P. (2006, May). Conflict resolution: what nurses need to know. Retrieved 17/12/2013 from: http://www.mediatecalm.ca/pdfs/what%20nurses%20need%20to%20know.pdf.

Mentis, H., O'Hara, K., Sellen, A., and Trivedi, R. (2012) Interaction Proxemics and Image use in Neurosurgery. *Proceedings of the SIGCHI Conference on Human Factors in Computing Systems* (CHI 2012), 927–936.

Mentis, H. M., Chellali, A., and Schwaitzberg, S. (2014). Learning to See the Body: Supporting Instructional Practices in Laparoscopic Surgical Procedures. In *Proceedings of the 32nd annual ACM conference on Human factors in computing systems*, New York, NY: ACM Press, 2113–2122. DOI: 10.1145/2556288.2557387.

Mentis, H.M. and Taylor, A.S. (2013). Imaging the Body: Embodied Vision in Minimally Invasive Surgery. In *Proceedings of the SIGCHI Conference on Human Factors in Computing Systems*, New York, NY: ACM Press, 1479–1488. DOI: 10.1145/2470654.2466197.

The Mid Staffordshire NHS Foundation Trust Public Inquiry (2013). Report of the Mid Staffordshire NHS Foundation Trust Public Inquiry. London: The Stationery Office.

Miles, M. B. and Huberman, A. M. (1994). *Qualitative Data Analysis: An Expanded Sourcebook*. Thousand Oaks, California, SAGE Publications.

Morrison , C., Walker, G., Ruggeri, K., and Hacker Hughes, J. (In press). An Implementation Pilot of the MindBalance Web-Based Intervention for Depression in Three IAPT Services. *The Cognitive Behavioural Therapist*. DOI:10.1017/S1754470X14000221.

NIHR RDS PPI Handbook (Patient and Public Involvement in Health and Social Care Research: A Handbook for Researchers by Research Design Service London). (n.d.). http://www.rdslondon.co.uk/RDSLondon/media/RDSContent/files/PDFs/RDS_PPI-Handbook_web.pdf. Accessed 18th February 2014.

Noy, C. (2008). Sampling Knowledge: The Hermeneutics of Snowball Sampling in Qualitative Research. *International Journal of Social Research Methodology*, 11(4), 327–344. DOI: 10.1080/13645570701401305.

Oakley, E., Stocker, S., Staubli, G., and Young, S. (2006). Using Video Recording to Identify Management Errors in Pediatric Trauma Resuscitation. *Pediatrics* 117(3), 658–664. DOI: 10.1542/peds.2004-1803.

O'Hara, K., Gonzalez, G., Penney, G., Sellen, A., Corish, R., Mentis, H., Varnavas A. et al. (2014). Interactional Order and Constructed Ways of Seeing with Touchless Imaging Systems in Surgery. *Computer Supported Cooperative Work*, 23(3), 299–337. DOI: 10.1007/s10606-014-9203-4.

O'Kane, A. A., Rogers, Y., and Blandford, A. (2014). Gaining Empathy For Non-Routine Mobile Device Use Through Autoethnography. In *Proceeding of CHI 2014*, ACM Press, 987-990. DOI: 10.1145/2556288.2557179.

Øvretveit, J. 1992. *Health Services Quality*. Blackwell, Oxford.

Pellatt, G. (2003). Ethnography and Reflexivity: Emotions and Feelings in Fieldwork. *Nurse Researcher*, 10(3), 28–37. DOI: 10.7748/nr2003.04.10.3.28.c5894.

Polson, P. G., Lewis, C., Rieman, J., and Wharton, C. (1992). Cognitive Walkthroughs: A Method for Theory-Based Evaluation of User Interfaces. *International Journal of man-machine studies*, 36(5), 741–773. DOI: 10.1016/0020-7373(92)90039-N.

Pronovost, P. J., Goeschel, C. A., Colantuoni, E., Watson, S., Lubomski, L. H., Berenholtz, S. M., ... and Needham, D. (2010). Sustaining Reductions in Catheter Related Bloodstream Infections in Michigan Intensive Care Units: Observational Study. *BMJ: British Medical Journal*, 340. DOI: 10.1136/bmj.c309.

Rajkomar, A. and Blandford, A. (2012). Understanding Infusion Administration in the ICU through Distributed Cognition. *Journal of Biomedical Informatics* 45(3), 580–90. DOI: 10.1016/j.jbi.2012.02.003.

Randell, R. (2003). User customisation of medical devices: the reality and the possibilities. Cognition, Technology & Work 5, 163–170. DOI: 10.1007/s10111-003-0124-0.

Randell. R., Wilson, S., Woodward, P., and Galliers, J. (2011). The ConStratO Model of Handover: A Tool to Support Technology Fesign and Rvaluation. *Behaviour and Information Technology*, 30(4), 489–498. DOI: 10.1080/0144929X.2010.547220.

Reddy, M.C. and Dourish, P. (2002). A Finger on the Pulse: Temporal Rhythms on Information Seeking in Medical Work. In *Proceedings from The Conference on Computer-Supported Cooperative Work*. New York, NY: ACM Press, 344–353. DOI: 10.1145/587078.587126.

Redfern, Michael, J. Keeling, and E. Powell. (2001). *The Report of the Royal Liverpool Children's Inquiry*. London: The Stationery Office.

Redfern, M. K. (2001). *The Royal Liverpool Children's Inquiry Report*. London: Stationery Office.

Research Governance Framework for Health and Social Care (RGFHSC) (Department of Health, 2005).

Ritchie, J. (2012). *The Application of Qualitative Methods to Social Research. Qualitative Research Practice: A Guide for Social Science Students and Researchers*. J. Ritchie and J. Lewis. London, SAGE Publications: 24–46.

Ritchie, J., Lewis, J., and Elam, J. (2012). Designing and Selecting Samples. Qualitative Research Practice: *A Guide for Social Science Students and Researchers*. J. Ritchie and J. Lewis. London, SAGE Publications: 77–108.

Rogers, C. R. (1962). The Interpersonal Relationship. *Harvard Educational Review*, 32(4), 416–429.

Sanders, P. (2002). *First Steps in Counselling*. Ross-on-Wye: PCCS Books.

Sarcevic, A., Marsic, I., Lesk, M. E., and Burd, R. S. (2008). Transactive Memory in Trauma Resuscitation. *Proceedings of the ACM 2008 Conference on Computer Supported Cooperative Work* (CSCW 2008), 215–224. DOI: 10.1145/1460563.1460597.

Sarcevic, A., Marsic, I., and Burd, R. S. (2012). Teamwork Errors in Trauma resuscitation. *ACM Transactions on Computer-Human Interaction* (TOCHI) 19(2), article 13. DOI: 10.1145/2240156.2240161.

Scupelli, P.G., Xiao, Y., Fussell, S.R., Kiesler, S., and Gross, M.D. (2010). Supporting Coordination in Surgical Suites: Physical Aspects of Common Information Spaces. In *Proceedings of the SIGCHI Conference on Human Factors in Computing Systems*, New York, NY: ACM Press, 1777–1786. DOI: 10.1145/1753326.1753593.

Sellen, K., Chignell, M., Straus, S., Callum, J., Pendergrast, J., and Halliday, A. (2010). Using Motion Sensing to Study Human Computer Interaction in Hospital Settings. In *Proceedings of Measuring Behavior* 2010, Eindhoven, The Netherlands, August 24–27, 2010. Eds. A.J. Spink, F. Grieco, O.E. Krips, L.W.S. Loijens, L.P.J.J. Noldus, and P.H. Zimmerman.

Sim, J. (1998). Collecting and Analysing Qualitative Data: Issues Raised by the Focus Group. *Journal of Advanced Nursing* 28(2), 345–352. DOI: 10.1046/j.1365-2648.1998.00692.x.

Smith, J. A., Flowers, P., and Larkin, M. (2009). *Interpretative Phenomenological Analysis: Theory, Method and Research*. London: Sage

Srigley, J. A., Furness, C. D., Baker, G. R., and Gardam, M. (2014). Quantification of the Hawthorne Effect in Hand Hygiene Compliance Monitoring Using an Electronic Monitoring System: A Retrospective Cohort Study. *BMJ Quality and Safety*, bmjqs 2014. DOI: 10.1136/bmjqs-2014-003080.

Staley K (2009). *Exploring Impact: Public Involvement in NHS, Public Health and Social Care Research*. Eastleigh: INVOLVE.

Steier, F. E. (1991). *Research and Reflexivity*. Sage Publications, Inc.

Stewart, D. (Ed.). (2011). Making the Difference: Actively Involving Patients, Carers and the Public. Patient and Public Involvement The Way Forward: Examples and Evidence from the Clinical Research Network. *NIHR*, May 2011.

Sudman, S. (1976). *Applied Sampling* (p. 249). New York: Academic Press.

Svensson, M. S., Heath, C., and Luff, P. (2007). Instrumental Action: The Timely Exchange of Implements During Surgical Operations. *Proceedings of the 10th European Conference on Computer Supported Cooperative Work* (ECSCW 2007), 41–60. DOI: 10.1007/978-1-84800-031-5_3.

Taneva, S. and Law, E. (2007). In and Out of the Hospital: The Hidden Interface of High Fidelity Research via RFID. In *Human-Computer Interaction–INTERACT 2007* (624–627). Springer Berlin Heidelberg. DOI: 10.1007/978-3-540-74796-3_62.

Tang, C. and Carpendale, S. (2007). An Observational Study on Information Flow During Nurses' Shift Change. In *Proceedings of the SIGCHI Conference on Human Factors in Computing Systems*, New York, NY: ACM Press, 219–208. DOI: 10.1007/978-3-540-74796-3_62.

Thompson, D. A., Kass, N., Holzmueller, C., Marsteller, J. A., Martinez, E. A., Gurses, A. P., ... and Pronovost, P. J. (2012). Variation in Local Institutional Review Board Evaluations of a Multicenter Patient Safety Study. *Journal for Healthcare Quality*, 34(4), 33–39. DOI: 10.1111/j.1945-1474.2011.00150.x.

Tillmann-Healy, L. M. and Kiesinger, C. E. (2001). Mirrors: Seeing Each Other and Ourselves through Fieldwork. In Gilbert, K. (Ed.). *The Emotional Nature of Qualitative Research*, 81–108. CRC Press. DOI: 10.1201/9781420039283.ch5.

Vitae (no date). The Vitae Researcher Development Framework. https://www.vitae.ac.uk/researchers-professional-development/about-the-vitae-researcher-development-framework/developing-the-vitae-researcher-development-framework, Accessed July 11, 2014.

Vincent, C., Moorthy, K., Sarker, S., Chang, A., and Darzi, A. (2004). System Approaches to Surgical Quality and Safety: From Concept to Measurement. *Annals of Surgery* 239, 475–482. DOI: 10.1097/01.sla.0000118753.22830.41.

Walton, M. M. (2006). Hierarchies: the Berlin Wall of Patient Safety. *Quality and Safety in Health Care*, 15(4), 229–230. DOI: 10.1136/qshc.2006.019240.

Watkins Jr, C. E. (1983). Burnout in Counseling Practice: Some Potential Professional and Personal Hazards of Becoming a Counselor. [Article]. *Personnel and Guidance Journal*, 61(5), 304. DOI: 10.1111/j.2164-4918.1983.tb00031.x.

Webb , T. L., Joseph, J., Yardley, L., and Michie, S. (2010). Using the Internet to Promote Health Behavior Change: A Systematic Review and Meta-Analysis of the Impact of Theoretical Basis, Use of Behavior Change Techniques, and Mode of Delivery on Efficacy. *Journal of Medical Internet Research*, 12 (1). DOI: 10.2196/jmir.1376.

Westbrook, J. I., Woods, A., Rob, M. I., Dunsmuir, W. T. M., and Day, R. O. (2010). Association of Interruptions with an Increased Risk and Severity of Medication Administration Errors. *Archives of Internal Medicine*, 170(8), 683–690. DOI:10.1001/archinternmed.2010.65.

WHO. (2013). *Ethical Issues in Patient Safety Research: Interpreting Existing Guidance*. World Health Organisation 2013.

Wilson, S., Randell, R., Galliers, J., and Woodward, P. (2009). Reconceptualising Clinical Handover: Information Sharing for Situation Awareness. *Proceedings of ECCE 2009—European Conference on Cognitive Ergonomics*, 315–322.

Wilson, S., Woodward, P., and Randell, R. (2010). PaperChain: A Collaborative Healthcare System Grounded in Field Study Work. In *Proceedings of the First International Workshop on Interactive Systems in Healthcare* (WISH), 177–180.

Wincup, E. (2001). Feminist Research with Women Awaiting Trial: The Effects on Participants in the Qualitative Research Process. In Gilbert, K. (Ed.). *The Emotional Nature of Qualitative Research*, 17–36. CRC Press. DOI: 10.1201/9781420039283.ch2.

World Medical Association (2008). WMA Declaration of Helsinki—Ethical Principles for Medical Research Involving Human Subjects. http://www.wma.net/en/30publications/10policies /b3/. Accessed 22nd July 2011. DOI: 10.1001/jama.2013.281053.

Xiao, Y. and Mackenzie, C. F. (2004). Introduction to the Special Issue on Video-Based Research in High-Risk Settings: Methodology and Experience. *Cognition, Technology and Work* 6(3), 127–130. DOI: 10.1007/s10111-004-0153-3.

Zheng, K., Padman, R., Johnson, M. P., and Diamond, H. S. (2009). An Interface-Driven Analysis of User Interactions with an Electronic Health Records System. *Journal of of American Medical Informatics Association* 16(2),228–37. DOI: 10.1197/jamia.M2852.

Zhou, X. (2010). Information in Healthcare: An Ethnographic Analysis of a Hospital Ward. (Doctoral dissertation). Retrieved from University of Michigan, http://deepblue.lib.umich.edu.

Zhou, X., Ackerman, M. S., and Zheng, K. (2009). I Just Don't Know Why It's Gone: Maintaining Informal Information use in Inpatient Care. *Proceedings of the 27th ACM Conference on Human Factors in Computing Systems* (CHI 2009), 2061–2070. DOI: 10.1145/1518701.1519014.

Zhou, X., Ackerman, M. S., and Zheng, K. (2010). Doctors and Psychosocial Information: Records and Reuse in Inpatient Care. *Proceedings of the 28th ACM Conference on Human Factors in Computing Systems* (CHI 2010), 1767–1776. DOI: 10.1145/1753326.1753592.

Biographies

EDITOR BIOGRAPHIES

Dominic Furniss is a Researcher Co-Investigator on the CHI+MED project at University College London. He investigates the design and use of medical devices in hospitals. His interests include the development of theory to support the understanding of performance in socio-technical systems. He is the lead editor and a co-author of Chapters 1, 2, and 6.

> UCL Interaction Centre, University College London, Gower Street, London WC1E 6BT, U.K.

Rebecca Randell is a Senior Translational Research Fellow in the School of Healthcare, University of Leeds, where she leads the decision-making research theme. Her research focuses on studying how technology impacts the decision making of healthcare professionals. She is a co-editor and a co-author of Chapters 1, 4, and 6.

> School of Healthcare, University of Leeds, Leeds LS2 9UT, U.K.

Aisling Ann O'Kane is a Ph.D. student on the CHI+MED project at University College London. Her research is on the situated use of mobile medical technologies for everyday self-management of chronic conditions. Her interests include the connections between human factors engineering and user experience in healthcare. She is a co-editor and a co-author of Chapters 1 and 3.

> UCL Interaction Centre, University College London, Gower Street, London WC1E 6BT, U.K.

Svetlena Taneva is a Human Factors Specialist at Healthcare Human Factors, University Health Network in Toronto. Svetlena specializes in the development and evaluation of technology and organisational processes for clinical environments. For the past 8 years, Svetlena has worked and published extensively in the area of HCI in healthcare. She is a co-editor for this book and also a co-author of Chapters 1, 3, and 6.

> Healthcare Human Factors, University Health Network, 190 Elizabeth Street, Toronto, ON M5G 2C4, Canada.

Helena Mentis is an Assistant Professor in the Department of Information Systems at the University of Maryland, Baltimore County. She examines the challenges clinical healthcare providers face in the embodied sharing and understanding of ambiguous and interpretive health information. She has conducted fieldwork in healthcare in the U.S. and the U.K. She is a co-editor and co-author of Chapters 1 and 6.

Department of Information Systems, University of Maryland, Baltimore County, 1000 Hilltop Circle, Baltimore, MD 21250, U.S.

Ann Blandford is Professor of Human–Computer Interaction at UCL, and leads the CHI+MED and ECLIPSE projects on making interactive medical devices safer. Her expertise is in models and methods for studying interactive systems "in the wild," with a particular focus on healthcare. She is the senior editor and a co-author of Chapters 1 and 6.

UCL Interaction Centre, University College London, Gower Street, London WC1E 6BT, U.K.

AUTHOR BIOGRAPHIES

Anjum Chagpar is the Managing Director of Healthcare Human Factors, a team of 25 engineers, psychologists, and designers transforming healthcare through the application of Human Factors. Her interests include the design of health technologies for under-served populations as well as emerging economies. She is a co-author of Chapters 2 and 6.

Healthcare Human Factors, University Health Network, 190 Elizabeth Street, Toronto, ON M5G2C4, Canada.

Deborah Chan is a Human Factors Engineer with Healthcare Human Factors, University Health Network. She has been designing user interfaces and evaluating applications for usability for over 10 years. Her focus is on the use human factors engineering approaches in healthcare. She is a co-author of Chapter 4.

University Health Network, R. Fraser Elliott Building, 190 Elizabeth St., Toronto, ON, M5G 2C4, Canada.

Yunan Chen is an Associate Professor in Informatics at the University of California, Irvine. Her recent projects explore the use of Health IT systems in various clinical settings, with a specific focus on clinical documentation, and patient-provider interaction. She is a co-author of Chapter 4.

Institute for Clinical and Translational Science, University of California, 5066 Bren Hall, Irvine, CA, 92697, U.S.

Andy Dearden is a Professor of Interactive Systems Design at Sheffield Hallam University, U.K. He leads research on participatory design and innovation with private, public, third sector, and grassroots community groups in the U.K., India, and Africa. He was the technical lead for the User Centred Healthcare Design project. He is a co-author of Chapters 1 and 5.

Cultural, Communication and Computing Research Institute, Sheffield Hallam University, Howard Street, Sheffield, S1 1WB, U.K.

Mads Frost is a Postdoctoral Fellow at the IT University of Copenhagen. His research revolves around the design and use of Personal Health Technologies in healthcare, focusing on pervasive and ubiquitous computing, quantified self, and data analysis as the primary research areas. He is a co-author of Chapter 5.

IT University of Copenhagen, Rued Laanggaards Vej 7, 2300 Copenhagen S, Denmark.

Kristina Groth is an Adjunct Professor in HCI at the Royal Institute of Technology and responsible for Telemedicine development at Karolinska University Hospital. Her research and development work focuses on computer-supported collaborative work and mediated communication. She is a co-author of Chapter 3.

Media Technology and Interaction Design,, Lindstedtsvägen 3, 100 44 Stockholm, Sweden.

Karolinska University Hospital, Innovation Centre, 141 86 Stockholm, Sweden.

Brian Hilligoss is an Assistant Professor in the College of Public Health at The Ohio State University. He investigates the interplay of organisational routines, communication, and information systems in healthcare. His interests focus on enhancing high reliability in complex socio-technical systems. He is a co-author of Chapter 3.

Health Services Management and Policy, College of Public Health, Ohio State University, 1841 Neil Avenue, Columbus, OH 43210, U.S.

Cecily Morrison is a Postdoctoral Researcher at Microsoft Research. Her research focuses on how people can be engaged with their health through digital systems, spanning idea generation to evaluation of systems in healthcare contexts. She is a co-author of Chapter 5.

Microsoft Research, 21 Station Rd, Cambridge, CB1 2FB, U.K.

Atish Rajkomar is a Ph.D. graduate of University College London. His research is on understanding patients' situated interactions with home haemodialysis machines. His interests include the design and evaluation of healthcare socio-technical systems, particularly using cognitive engineering approaches. He is a co-author of Chapter 4.

UCL Interaction Centre, University College London, Gower Street, London WC1E 6BT, U.K.

Raj Ratwani is the Scientific Director for MedStar Health's National Center for Human Factors in Healthcare and an Assistant Professor of Emergency Medicine at the Georgetown University School of Medicine. He is the primary investigator on several grants focused on applying human factors principles to advance health and is an expert in the areas of memory and perception. He is a co-author of Chapters 3 and 6.

National Center for Human Factors in Healthcare, MedStar Health, Washington, D.C., U.S.

Aleksandra Sarcevic is an Assistant Professor at the College of Computing and Informatics at Drexel University. Her research interests are in computer-supported cooperative work, focusing on ethnographic studies of practice and coordination in high-risk medical settings that inform technology design and implementation. She is a co-author of Chapter 4.

College of Computing and Informatics, Drexel University, 3141 Chestnut St., Philadelphia, PA 19104, U.S.

Katherine Sellen is an Assistant Professor at OCAD University where she teaches human factors. She is interested in understanding adoption and adaptation behaviors to new technology in clinical settings, innovation approaches in healthcare, and the role of design in reducing medical error. She is a co-author of Chapter 4.

Faculty of Design, OCAD University, 205 Richmond St., Toronto, ON, Canada.

Anja Thieme is a Postdoctoral Researcher in Culture Lab at Newcastle University. Her research focuses on sensitive and empathic approaches in the design and evaluation of technology to promote the mental health and wellbeing of people who have complex mental health problems. She is a co-author of Chapter 3.

Culture Lab, Newcastle University, King's Walk, Newcastle upon Tyne, NE1 7RU, U.K.

Ross Thomson is a Ph.D. student on the MATCH project at the University of Nottingham. His research is on home use medical devices and older people. His interests include using interpretative phenomenological methods to describe the psychosocial impact of medical technologies. He is a co-author of Chapter 2.

Faculty of Engineering, University of Nottingham, Nottingham, NG7 2RD, U.K.

Heather Underwood is an Assistant Professor at the University of Colorado Denver. Her research focuses on health informatics, maternal and public health point-of-care technologies, and information and communication technologies (ICTs) for the developing world. She is a co-author of Chapter 2.

University of Colorado Boulder, ATLAS Institute, Colorado, U.S.

Daniel Wolstenholme is a Clinical Researcher at Sheffield Teaching Hospitals NHS Foundation Trust and Visiting Researcher at Sheffield Hallam University Art and Design Research Centre. His current research is specifically around the application of design theory and practice in health and social care settings. He is a co-author of Chapters 1 and 5.

Sheffield Teaching Hospitals NHS Foundation Trust, Royal Hallamshire Hospital, Glossop Road, Sheffield, S10 2JF, U.K.

Xiaomu Zhou is an assistant professor in the School of Communication and Information at Rutgers, The State University of New Jersey (New Brunswick). She examines the clinical documentation, patient-provider communication, and Electronic Health Records implementation and adoption. She is a co-author of Chapter 4.

School of Communication and Information, Rutgers, The State University of New Jersey, 4 Huntington Street, New Brunswick, NJ 08901, U.S.

Fieldwork for Healthcare: Volume 1